A Platform
Mindset

A Platform Mindset

/ My lessons from
developer to CTO

< **MARCUS FONTOURA**

8080
BOOKS

an imprint of Microsoft

Published in the United States by 8080 Books,
an imprint of Microsoft Corporation
1 Microsoft Way, Redmond, WA 98052

https://aka.ms/8080books

Cover Design by Shyla Lindsey
Typesetting by Mariner Media
In association with Partners Media

ISBN 979-8-9991201-6-8 (Paperback – 8080 Books)
ISBN 979-8-9997550-1-8 (Hard Cover – 8080 Books)
ISBN 979-8-9997550-0-1 (eBook – 8080 Books)

MICROSOFT, 8080 BOOKS, and the 8080 BOOKS Logo
are trademarks of Microsoft Corporation

To my daughters, Ayesha and Shanti,
and to my parents, Vanessa and Julio.

Table of Contents

PART III: LEADERSHIP

"Marcus Fontoura's *Platform Mindset* avoids the tiresome rhetoric of yesterday's leadership and management books to focus on what matters most in this digital age—supportive, inclusive work environments where creativity and innovation flourish to solve complex challenges for customers and the world."

Satya Nadella
CEO Microsoft

"Marcus Fontoura's *Platform Mindset* teaches us how to build inclusive organizations where people and teams thrive. Full of engaging stories and best practices, this book is a great read for anyone working or considering a career in technology! More than that, this book is a story of a leader learning how to architect a high-performing team and culture. Having learned from some of the best in technology, Marcus takes us on an exciting ride with some great views from inside the ropes."

Christine Porath
author *Mastering Civility: A Manifesto for the Workplace*

"Marcus is an incredible technologist who has been at the forefront of many industry-shaping technology innovations, from computational advertising to cloud computing to fintech. Through his book, Marcus takes us along on his personal journey of learning, insights, and growth as a tech innovator. This is a must read for anyone who is interested in gaining a deeper understanding of the inner workings of technology management."

David Ku
entrepreneur, former Corporate Vice President at Microsoft and Senior Vice President at Yahoo

"Marcus approaches the software development lifecycle in a tech environment with ease, given his extensive experience in the field. This is a valuable and useful read for anyone wishing to make technology a central part of their business, as a fundamental component for achieving scalability and efficiency that cannot be obtained through alternative means."

Berthier Ribeiro-Neto
co-author of *Modern Information Retrieval*

"With a narrative rich in personal and professional experiences, Marcus offers a unique and practical perspective on how to lead and innovate in a dynamic technological environment. His ability to translate complex technological concepts into accessible business contexts makes this book an essential read for leaders and aspiring tech managers. Happy reading!"

Paulo Caroli
speaker, consultant, and author of *Lean Inception: How to Align People and Build the Right Product*

"I believe the reader of this book will benefit in two ways: first, from the technology and management content the book covers, which is its primary objective; and second, from learning how it was possible to achieve such a level of knowledge and expertise in computer science here in Brazil—knowledge with global relevance, built by a few individuals through their own effort and many years of dedication."

Prof. Cesar Simões Salim
Pontifical Catholic University of Rio de Janeiro
(PUC-Rio)

"In *Platform Mindset*, Marcus Fontoura distills the unassuming wisdom gained from a remarkable career at some of Silicon Valley's most influential companies. With unparalleled clarity, Marcus guides readers through the essential elements—culture, technology, and teamwork—that drive organizations to redefine what's possible in an era of rapid technological advancement. Drawing from his own transformative contributions to groundbreaking platforms, Marcus provides an insightful roadmap for leaders aspiring to build companies capable of expanding the boundaries of innovation. This book is an indispensable guide for anyone seeking to create technology that not only succeeds but reshapes the future."

Vanja Josifovski
CEO and founder, Kumo.AI

Foreword

It is with great pleasure that I present the book *Platform Mindset* by Marcus Fontoura. This work offers a practical and reflective perspective on technology management, based on Marcus's vast experience over the last 20 years working in the US at companies such as IBM, Yahoo, Google, Microsoft, and most recently at Stone, where he has played a crucial role in the company's technological transformation.

From an early age, Marcus faced challenges that shaped his unique approach to technology and management. He didn't excel in traditional subjects but found a passion for mathematics that guided him to success. This challenging start and his determination to overcome difficulties reflect the essence of his professional journey.

In this book, Marcus shares his experiences and lessons learned, providing a detailed perspective on how technology should be conceived and executed to drive businesses forward. He addresses crucial aspects for leaders and managers, discussing how to attract and retain exceptional talent, the importance of standardized systems and engineering tools, and how to handle inevitable technical challenges with a distinctly human touch.

Platform Mindset is divided into three fundamental parts: Culture, Technology, and Leadership. In the section on Culture, Marcus discusses the importance of creating an environment that values innovation and creativity, attracting the best talent and keeping them motivated, allowing for mistakes not to be punished, but rather encouraging a true culture of experimentation, where failure is just the beginning of a discovery process. He also offers practical guidance on how to hire, develop, and manage an efficient and diverse technology team. At Stone, these practices have been essential in transforming the company from a startup in 2012 to a cutting-edge technology powerhouse.

The Technology section covers the construction of standardized platforms that increase productivity and foster collaboration among teams. Marcus explains the relevance of consistent and uniform processes and the importance of code review as an educational tool for the team.

Regarding Leadership, the author emphasizes that creativity and ideas are irreplaceable assets. He discusses diversity in the corporate environment and how a technology team can have a significant social impact, promoting education and transformation both within and outside the company. At Stone, Marcus's leadership has been fundamental in fostering a culture of innovation and collaboration, which is key to the continued success of any organization.

With a solid foundation in real examples and personal stories, Marcus invites us to rethink technology management, bringing valuable insights to professionals at all levels. I am confident that reading

Platform Mindset will enrich your understanding of technology management and inspire new approaches in your career and personal life.

Happy reading!

Andre Street
Founder of Stone

Preface

In 2013, I was working as a research scientist and developer at Microsoft. Once a competitive, revered brand in personal computers, the company had fallen behind the industry after basking in the grandeur of the Windows 95 launch. Amazon was leading the way in cloud computing and Google was climbing the summit of search. We were passionate but we were also a distant player in both cloud and search.

Yet here I was, a lowly individual contributor, with a remit to entice corporate vice presidents to abandon their individual fiefdoms in favor of building a more unified, customer-first democracy. A rising star inside the company, Satya Nadella, had run Bing and was now preaching a gospel we called, "One Microsoft" as we worked to bootstrap the business that would become Azure.

In those days, Bing, Azure, Exchange, Windows, and others were more like loose affiliations of software capabilities. They were not yet the integrated platform they would become. To get there, it required solving hard technical problems, breaking down organizational barriers, and building a new culture of collaboration.

This was not just a nice idea—it was absolutely urgent. Far behind Amazon Web Services (AWS), Azure needed to become a viable alternative for businesses ranging from digital-native startups to giants in manufacturing, financial services, health care, government, and the like. As balkanized brands, we lacked the flexibility to handle massive compute demands without a unified platform approach.

Satya set the tone by creating the expectation that everyone's individual evaluations would be based on how each person contributed to the success of others. That cultural mission inspired me to use my technical skills to clarify a cross-organization, pan-product vision. With the right cultural and technical roadmap, every individual fiefdom would see the obvious benefits of unification. Microsoft would not have to demand that everyone set down their guns; everyone would simply see the inevitable intelligence of joining together to create something bigger and better for customers.

Even as I write this, I realize it sounds a little too simple. The truth is that it was very hard and took time. What I realize now is that Microsoft's leadership understood this. We needed to get it right. We needed to create the right environment for newbies and veterans to grasp what was possible and do the right thing.

My career, and indeed this book, has largely been about the ethos needed to build powerful platforms. So what is a platform? I define it as a reusable system that implements a core piece of functionality, such as an identity platform, which provides a solution for users to log-in to multiple services. Developers gain agility when using platforms, as they don't need to implement the whole system from scratch and can focus on the new, differentiating innovations. However, a platform is an external dependency, and using a platform implies teams taking a dependency on other teams. This requires coordination and collaboration. It is impossible to build the large-scale, complex systems of today without relying on platforms and their multiplier effect—build a scalable identity solution once and reuse it many times in all the company's services.

This means that modern technology organizations need to be highly collaborative, with all their sub-teams interacting in an open way, in which ideas are generated, developed, and propagated in an environment that empowers the developers to be themselves and let their creativity flourish. Platforms are the seeds of innovation and require their daily dose of creativity and collaboration to blossom into the systems that impact our lives and society in a positive way. I truly believe we can create a better world through technology and a platform mindset is the foundation.

This book is about that, and its lessons are relevant for leaders in every industry and sector.

◆◆◆

On December 2023 I interviewed Microsoft CEO Satya Nadella for Stone Camp, Stone's end of year leadership event. In that interview, I asked him about his views on leadership and culture, and his experience leading the Microsoft cultural transformation. Satya's work, which I experienced first-hand during my tenure there, relied on Carol Dweck's **growth mindset**[1] ideas and scaled it to the whole organization, making Microsoft more collaborative and unlocking huge value from all the innovated work that followed. Leaders from different parts of the organization started working together and new synergies emerged.

At the time of the interview, I was already thinking about a book that would help other engineers and leaders understand the job of a CTO, including all of its different facets—at least all the aspects that I consider relevant—and I wanted to ground this book in my practical experience through stories of my career of 20+ years in big tech companies, starting as a developer and researcher and moving through the ranks.

I liked the word "mindset" because I wrote this book from the perspective of the leader—what he or she has to do. And it starts from a growth mindset. I didn't know this term and Carol Dweck's work until very late in my career, but I always believed that being willing to learn is what allowed me to grow as a developer and do very different things, such as being the chief architect for Azure compute and leading engineering organizations of 1,500+ people.

---

1. The growth mindset expressed in the book *Mindset: The New Psychology of Success*, by Carol S. Dweck (Random House, 2007), is one of the foundations of the cultural transformation led by Satya Nadella at Microsoft, which I was able to witness at close quarters.

Platform mindset is my attempt to systematize the work of the engineering leader—focusing on culture, technology, and leadership. The core of it is how to create an environment that fosters collaboration and potentializes innovation. Collaboration is core, as all the innovation that has the potential to impact the world is complex and cannot be done alone—even if you consider yourself really smart. We can do more together, and culture, technology, and leadership are the ingredients that will make it happen.

Introduction

I remember the reaction at our home when my parents received a letter from the school telling them that all my grades were bad. I was in 6th grade, and I really wasn't a good student. The world of rote learning, which is basically how we all were taught at the time, didn't interest me. It also seemed to present an obstacle to my dyslexic brain, which processed written text in a different way. I didn't think I could keep up with information in that stagnant way, and I ended up not trusting my ability to learn because of the problems I had with reading aloud. I used to get very nervous when teachers asked me to read to the class. Words on paper would jumble up and disappear in my head before the next sentence even arrived, my voice would become uncertain, and the cadence of my speech would be lost, resulting in a total mismatch between my thoughts and the way I expressed myself.

My father, Julio Cezar, a chemical engineer at Petrobras, a state-owned oil company in Rio de Janeiro, and as a typical Brazilian patriarch of the 1980s, was always away working and took little part in the family's day-to-day life. Contrary to my mother, Vanessa, who juggled completing her degree in psychoanalysis and opening her practice while taking care of her family and was always very present and caring. They met when they were still children living on the same street in Fortaleza, in the northeast of Brazil, married early, moved to different cities for his work, and eventually settled in Rio de Janeiro. There is where my sister, Tatiana, was born, and where we were raised. We lived a middle-class life with no luxury, but they always made our education their economic priority. So, on the day that letter came from school, my father was furious, probably ashamed, and declared: "From now on, you're going to study three hours every day at home after school to get your grades back up."

I don't know if it was because my father wasn't very present in our lives and we didn't interact that much, but I always listened to everything he said, and his words always had a big impact on me. Following my father's orders and those of the private math tutor they had arranged for me, I stacked some books and began to use the three hours I spent sitting at the desk in my room to study.

I didn't become an exceptional student in all subjects, but something in those Algebra I books began to intrigue and satisfy me. I did all its exercises once, then did them all again and started devouring other math books that intrigued me. Unlike long history texts, which seemed intangible to me at the time, math exercises provided an objective result with a right or wrong answer, and an answer key at the end to check my answers immediately. As I didn't have permission to leave my room and was a bit uncoordinated when it came to sports, I stayed studying and gradually began to understand mathematics. More than that: I fell in love with mathematics and number theory, which is the area of mathematics that studies the properties of numbers, especially positive integers.[2] I loved solving problems involving factorization and would often spend the whole night working on them. Over time, I got my grades up and had a turnaround as a student in the following years. In the first year of high school, I was already a very good student in math and physics and joined the prep class for the Olympiads in these subjects, which I participated in without, however, achieving incredible results.

Let it be said that I didn't discover myself to be a naturally extraordinary mathematician. I dedicated a lot of effort to it and put in the discipline, organization, and patience to make clearer a lot of things that didn't make sense to me. I studied hard, and continued to do so ever since, to learn new things and challenge myself to go beyond my current viewpoints and limitations. I didn't know about *growth mindset* at the time, but that is how I started to think and operate. Through the years, I've also realized how valuable it is to surround myself with talented, inspiring people who are smarter than me, both in academic and professional environments.

This may seem like an exercise in humility, and to a certain extent it is, but it has served me as a form of self-knowledge. I'm aware that I'm not the most

---

2. Number theory has many practical uses, cryptography and digital security amongst them.

brilliant person. I'm not going to bring about the next revolution in technology or customs. However, with my own maturity, I have also understood the things that I *can* do very well: work across research and product development, translating the theoretical knowledge and vocabulary to implementation, and coordinating development and research teams. I found my professional place on this intersection between numbers, systems, and the synergy of people and their varied—and not always evidently complementary—knowledge.

If this pragmatism involves some reflection on my limits and skills, it certainly involves an even bigger dose of observation about how colleagues, managers, advisors, and mentors I've encountered throughout my life thought and acted. By observing them, I came to the conviction that some people are fundamental to a business's success.

In this book I share my experience of more than two decades working at IBM, Yahoo, Google, and Microsoft, where I witnessed the flourishing of technologies such as search engines and cloud computing from the inside. I discuss how to bring the spirit of big tech to companies, large or small, from the perspective of a technology leader. To accomplish this, technology should be designed and executed in a creative way and occupy a central role in the organizational decisions of basically any company wanting to expand and scale its processes. Only then would there be room for innovations that bring competitive advantage.

Many of the ideas presented here, including how to build and maintain a high-performance team, how to enable innovation, and how to develop reusable platforms are being put into practice as we speak at Stone, a Brazilian payments and financial services company that has been investing heavily to transform from its startup origins to a big tech. I wrote this book as its CTO (Chief Technology Officer) to also share a little of Stone's transformation story.

Organization of the book

Platform Mindset[3] was written for people in leadership roles, specifically those interested in learning more about technology and ways to use it to boost their business. This book can be useful for technology managers, from those at the helm of entire organizations to those running smaller projects and teams. But it is also helpful for managers in general, from all corporate areas, who want to gain a deeper understanding of how technology can shore up processes and make a transformational difference. And, I hope, it will be interesting for everyone who is just starting out in

---

3. This version of the book is an adaptation from the Portuguese version, *Tecnologia Intencional*, with a few important modifications, one of the main ones being the title, which was changed from "technology" to "platform," as platforms are the unifying concept in the book, and from "intentional" to "mindset" to highlight the importance of Carol Dweck's growth mindset work to the leadership principles proposed here.

business: students, developers, and managers at the beginning of their careers, who may be able to absorb the teachings of the incredible people I met and the style of leadership they planted in me, which I plan to describe here.

The book is divided into three parts:

- Part I—Culture (how to think about people, teams, and innovation)
- Part II—Technology (how to think about systems)
- Part III—Leadership (how to think about leaders)

Although the book's structure follows a logical order, presenting my management philosophy sequentially, each chapter can be read independently or consulted out of order. As happened at the beginning of this introduction, stories and examples from my personal and professional life will appear to complement the management principles being described.

Part I—Culture

I'll start by defining some basic concepts in Chapter 1; readers familiar with technology can skip this chapter. Here I'll share my vision of what it means to be a technology company and which role a leader should play in a truly digital environment. I'll also discuss how to think about technology in an intentional way, which I call the **platform mindset**, as it builds on the idea of *growth mindset* as a foundation of how to think about technology with intentionality and curiosity. I also talk about the cultural obstacles that can make it difficult to transform an organization into one where technology plays a central role.

I'm a great believer in the power of mentors and inspiring people. In Chapter 2, I'll discuss more broadly how to attract the most talented people for a role. I'll talk about how to keep them motivated, giving practical guidance on how to hire, evaluate, and develop a technology team that is efficient, diverse, and engaged, with room for professional growth in both management and technical areas.

Building and managing an efficient team is about more than just hiring well and skillfully developing individual professionals. Chapter 3 will delve into ways of structuring the team, including, for example, how to manage the seniority distribution of employees from different teams, how to manage individual contributors (also known as Staff+[4]) and how to organize hierarchical levels so that there is communication, agility, and freedom.

From there, I will explore the various ways of organizing a company to foster innovation in technology. I will discuss how, in the real world of competitiveness and tight budgets, we can create an environment that promotes freedom for the technol-

---

4. *Staff*+ is a term used to refer to individual contributors who are senior specialists and take part of the technical leadership of the organization, as described in the book *The Staff Engineer's Path*, by Tanya Reilly (O'Reilly Media, 2022).

ogy team to function as a creative organization. After all, it is creativity that generates truly innovative services and products.

Part II—Technology

With the team fundamentals in place, it's time to get down to business. In Chapter 5, I'll talk about building standardized platforms, which increase the productivity of the technology team by promoting collaboration. The creation of frameworks ensures that everyone builds upon the same foundations. This is not a ready-made formula: each company must create the solution best suited to its needs. To discuss this, I'll bring examples from the companies I've worked for.

In Chapter 6, I'll cover the subject of engineering tools as a way of implementing consistency and promoting collaboration. I'll talk about the importance of the development lifecycle, highlighting the importance of code review as a vehicle for team education. Engineering processes and tools have a huge impact on how teams collaborate and are a fundamental pillar, often overlooked by leaders, in shaping engineering culture. That's why my focus here will be on something other than the tools *per se*, but on their impact in collaboration and culture.

In technology, stumbles are to be expected. You can only innovate by making mistakes. In addition to human failures, equipment stops working and unforeseen events happen, no matter how prepared we are. In

Chapter 7, I explore technology incidents: why they happen, how to prevent them, and how to deal with them when they do occur.

Part III—Leadership

This is a book about technology in which the protagonist is not the machine but the human element. Creativity and ideas are irreplaceable assets. Fostering creativity involves promoting dialog between different people. The mix of worldviews, backgrounds, personal experiences, and diverse cultural histories within companies is one of the secret ingredients for the emergence of truly innovative solutions.

In Chapter 8, I will discuss diversity and inclusion (D&I) beyond the inclusion policies that "look good" in companies' sustainability reports. I see D&I as intrinsic to the principles of technology management. Beyond having a diverse group of people, they need to feel fully comfortable creating and operating in an environment that is respectful and free of micro-aggressions.

A company's technology team is a relevant social actor. It has the power to change the lives of employees and collaborators and can promote education both inside and outside the company. Chapter 9 explores the social role of technology as a tool for transformation, both in the microcosm of the team and more broadly. It was also the motivation for my return to the Brazilian market after more than twenty years working in big tech companies in the United States.

Finally, in the tenth and last chapter, I'll outline the attributes I believe a leader should strive to develop to put into practice everything I've described in the previous chapters—always conscious that there are no magic solutions because, unfortunately, they don't exist. With this book, I hope to pass on at least a little of what my mentors and collaborators taught me throughout my career: a style of leadership that respects individuality and fosters transformative ideas that bring about not only profits, but also solutions to facilitate exchanges between people around the world, and what all that these interactions have the potential to create.

PART I
CULTURE

Chapter 1
Basic concepts

When I joined Stone as CTO in 2022, I soon realized that the company had many complex systems and that technology had been a competitive differentiator since its inception. Stone's payment system, for example, is one of the most scalable systems in the world. However, several separate organizational areas were developing systems within the company, and these areas were not fully integrated, generating economic and technological losses.

The company's systems were developed with focus on the time-to-market—optimizing the time it takes for a product to reach its customers—which makes sense for a start-up in its hyper-growth phase. There was no CTO or technology leader to define guidelines, platforms, and architectural standards. What's more, other sectors of the company saw the technology area as a service provider rather than as a partner team that works together to build joint solutions.

In this chapter, I'll define some basic concepts that will be explored in depth in the rest of the book. I'll talk about what it takes to shift from a company that develops and uses technology internally, even on a large scale, to a **technology company** that views all its processes from a technological perspective. We'll also see that scale is one of the main reasons it's necessary to design products and processes from this perspective, as it's impossible to scale non-integrated systems and processes.

I will also define what it means to think **intentionally** about technology, striving for a conscious balance between outsourcing and developing systems and services in-house, between evolving existing products and innovating, and between technical debt and productivity. Sometimes, the transformation from a startup to a technology company that makes intentional decisions requires a cultural transformation—in general, it is easier to deal with systems than with people. Defining and working on the company's culture is a daily exercise that is also part of the technology leader's tasks.

Technology company

It's natural to associate the idea of a technology company with giants like Google, Meta, Apple, and Microsoft or with substantial scientific innovation ventures like NASA or the biotechnology consortiums responsible for the rapid development of COVID-19 vaccines. Beyond these, however, countless companies around the world create advanced technology and for many others technology permeates every aspect of their business.

Regardless of the industry or even its size, in a technology company all processes are evaluated carefully, in an intentional way, and with a learning mentality, and the entire company eventually operates through computer systems. And this isn't just about adopting digital processes or automation. Above all, it's working on day-to-day activities and defining corporate priorities, considering how technology will make a difference to the business now and in the future—especially when the company's growth creates the need to scale and diversify its products and services.

You must remember, of course, that a family-run micro-business may not yet fit the "technology company" category and may not even want to. However, if the business prospers and expands significantly, growing to 30, then 300, and even 1,000 people, it will become inevitable to consider, at some point, that only with technology as an integral part of the business will the company be able to improve its delivery network, receive and make payments, keep supplier and customer databases up to date, match payroll with employee data, and account for income and expenses.

In a company that follows the business-integrated model of platform mindset, a CTO or engineering leader is in charge of a team of developers with support functions for all the company's departments, from marketing to HR and sales. The HR manager knows that he has a counterpart in technology who is totally dedicated to HR and is curious to learn about the business and work in partnership to develop digital solutions.

Technology in the foreground

It's more than just small businesses that at some point need to re-evaluate their relationship with technology and overcome the belief that having an IT department is enough to solve technical problems as they arise. The same goes for medium-sized and large companies, corporations, and even governments, which can be at financial and reputational risk if their software is not secure and designed with internal and external dependencies in mind, and does not scale. Think of what usually happens when there is a lot of simultaneous access to a free product distribution website or mandatory registration on the last day of the deadline (or if you are a Swiftie and want concert tickets). If the system isn't built with this kind of scale in mind, it will inevitably crash.

A poorly designed system can mean anything, from lost purchases in a jewelry store because the cards didn't go through to a patient's data not being accessible to hospital staff or millions in losses if the breakdown is widespread in a banking or retail platform. The proper coordination of technology in a public or private institution, be it tiny or huge, is vital in our world.

It is all too common for organizations to perpetuate a disorganized process of managing technology, which tends to generate major losses over time—economic, technical, and human resources. This is even more common in companies that have grown too fast by using technical improvisation, also known as hacks, or unsustainable solutions built in-house or acquired from other companies.

Growing companies behave as if they were in a race against their competitors to get up to speed, meet deadlines, and launch products and services. This doesn't allow them to plan for the long term, and this lack of strategic thinking often puts their survival at stake.

In a technology environment without specific and well-thought-out management, there is no room for optimizing existing procedures and little chance of creative and innovative ideas that could represent a real competitive leap for the company. Larry Page, founder and former CEO of Google, used to say that we often need radical changes rather than incremental ones.[5] That is only possible if we intentionally think about innovation. To avoid cycles that risk the very existence of the business, it would be ideal for the company to understand its momentum and expansion potential and invest in technology management at early stages—in the same way, for example, that it may decide to bring in a human resources management professional when the number of employees exceeds a certain threshold.

However, as the ideal way is almost never the way the world works, it's more realistic to assume that

---

5. "Especially in technology, we need revolutionary change, not incremental change." Larry Page in an interview with *Wired* magazine in March 2014, available at https://www.wired.com/2014/03/larry-page-using-google-build-future-well-living/ (accessed Mar. 27, 2024).

companies will adopt technology management when there is already some level of disorganization and a lot to fix. The sooner the platform mindset is at the heart of decisions in all divisions, the better.

And how do you take this from theory to practice? The first step would be to structure the technology area with well-defined leadership, culture, systems, and processes. For a technology team to work well, it requires balance: between professionals with varying levels of experience, between managers and individual contributors; between time dedicated to the day-to-day "bread and butter" and time dedicated to innovation; between the necessity of meeting deadlines and the necessity of leaving some free time to creativity; and finding the middle ground between people with diverse backgrounds and different ways of thinking. Achieving these balances is no simple task, and each company tends to come up with its solution.

Regardless of the path between the ideal and the real world, it is essential to invest in a technology culture that can be accepted and implemented within all departments of the organization and to have a technology team capable of developing integrated solutions working in partnership with them.

Once this is achieved, the company shouldn't adopt a new sales channel, a logistics overhaul, or a people assessment system without involving the technology team. Technology becomes part of all these processes, not because the technology team is looking to boss around other areas of the company—but because all these processes would be cocreated and supported by the technology team.

In companies without a technology department properly involved in its processes, it is common for systems to be developed or purchased with no participation of the right people for the job and no clear guidelines. These decisions may work temporarily, but they create huge problems in the long term because it's harder to fix a system when you haven't followed its implementation step by step. With a proper platform mindset, the engineering processes are carried out correctly from the start. For example, using agile development methodologies, new systems can be validated gradually, starting with MVPs (minimum viable products[6]). This saves time and increases productivity.

How to deal with scale and technical debt

It's difficult to predict whether a company will ever grow, since most new businesses don't flourish in real life. You can be aware that, if it does grow, you'll need to think fast about solving the challenges in each area with an eye to scale, because artisanal solutions tend to be chaotic to manage in the long term. And mistakes in a small company have very different con-

---

6. A term used by companies or ventures to describe the most basic and least resource-intensive product in terms of time and money that can be launched. There are various methodologies for this, such as Lean Inception, described in the book *Lean Inception: How to align people and build the right product*, by Paulo Caroli (Editora Caroli, 2018).

sequences than mistakes in companies that already operate at scale. The best way to minimize mistakes is to be intentional about how to grow.

It's a risk not paying attention to the fact that the company got suddenly bigger. Still, each team continues to make independent decisions and follow the same processes as when the company was a start-up. For example, an employee performance review system that could handle 100 employees may not work so well when that number reaches 1,000.

Why not? Perhaps Alex from HR would copy the names of the people from an employee registration spreadsheet into the review system, circulate the information between the involved parties, then manually transfer the reviews and comments back from the review system to the spreadsheet. However, with a much larger number of people, departments, and responses involved, this kind of workflow becomes burdensome and eventually becomes a **technical debt**.

The lack of integration between the employee registration and the performance review systems would be nightmare for the managers. The list of employees to be evaluated would never be correct, as it would depend on manual processes that are susceptible to error and not scalable.

Without a platform mindset permeating the company systems and processes, technical debts become obstacles to growth, and the operation is compromised—even if the product or service is conceptually great. The accumulation of technical debt leads to a loss of productivity over time. Figure 1 shows that building software with shortcuts without attention to quality and the best way to implement each new feature can lead to an initial productivity gain. However, as would happen with a monetary debt, in the long term, the interest accrues, and the technical debts significantly impact productivity.

Human resources processes scale with the number of employees; sales processes scale with the number of clients; operations processes scale with the number of suppliers; legal processes scale with the number of employees, clients, and offices. And logistics problems scale with the number of headquarters, clients, and employees. Many aspects take on a large

Figure 1: Loss of productivity due to technical debts

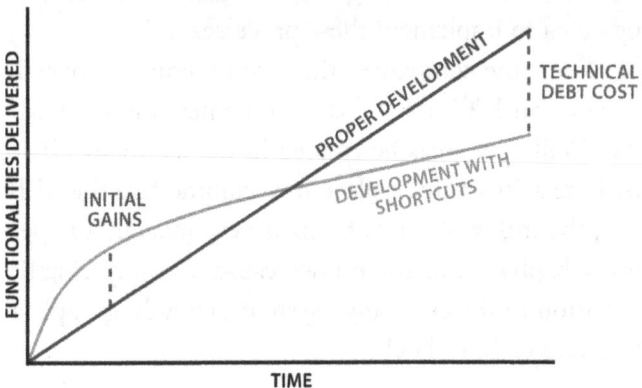

Source: Adapted from "Technical Debt - the silent villain of web development," by Piotr Golofit, on the Accesto blog, available at https://accesto.com/blog/technical-debt-the-silent-villain-of-web-development/ (accessed Mar. 27, 2024)

scale as a company grows, and to meet this inexorable demand, the people in leadership roles must turn their minds to building an **integrated technology area**.

It shouldn't be news to anyone that, in today's world, technology is a way of creating efficiencies in almost every process in manufacturing, industry, commerce, or services. This may not be the case for your business, but you can't reach this conclusion without first conducting an analysis along the lines of "Dodo, I really do not need a platform mindset to..." This reflection is even more relevant for those preparing to expand.

Each company grows differently and has different needs from the moment it starts scaling up. Managers need to consider how they want their processes to work and how they are going to organize the technology area to implement these processes.

For some companies, the turning point in growth is when an IPO occurs,[7] due to the new influx of capital. Will new areas be created in the company? If so, is there a structure in place to accommodate them? If not, the influx of funds to open new areas in a hypergrowth phase can sometimes cause a highly chaotic situation in the company's systems, including duplication and technical debt.

---

7. IPO stands for "initial public offering." This is when a company's shares are offered to the public on a stock exchange.

A dog food business that started out as a home-based business and suddenly became popular and needed to scale may have different priorities when it grows. It may need to open a better channel for active communication with customers, requiring specialized software. As the business grows, technology needs to change to support its processes and operations in an intentional and integrated way with the rest of the organization.

More and more options are emerging to support this expansion without an in-house technology department, such as contracting third-party services using cloud-based SAAS (software-as-a-service) solutions. Depending on the size of the company and its plans, this can end up becoming an effective permanent solution. It can also be a useful alternative for a few years until the process becomes more complex and the priorities or the business model itself change.

When technology managers deal with several technology service providers, internal developers, and a lot of clutter, it's time to rationalize the tech investments.

What is a platform mindset?

Investment in technology should generate value for the company. A large part of this investment goes into human resources because technology professionals tend to have higher salaries than other areas. Their potential to generate value is also very high, so a large part of the technology leader's job is to hand-pick the team.

Once you have a team that is as lean and competent as possible, you need to analyze what should be developed and managed in-house and what should be outsourced. To do this, the technology manager must determine which projects are strategic and should take the team's time, as well as which ones are commodities and should be outsourced.

This type of decision includes considering costs for both scenarios, supplier costs and internal costs, since even if a project is outsourced, it still must be monitored by the internal team and somehow integrated with the other company's systems.

Having a senior technology professional on board as early as possible is key to carry out these assessments. A technology manager who plays a leading role in a company's early stages would be able to plan all the processes, both from the point of view of systems development and implementation and in building the team and defining management processes.

On the other hand, a technology manager hired by a company that has grown disorganized is left with the more laborious task of reviewing all the processes and, in most cases, revisiting the configuration of the team itself. They will have to deal with questions such as: does the investment in technology pay off? Are people in the right roles? Is there a good ratio of senior and junior positions? Are they working on projects that make sense for the company's mission? Are there any redundant projects in the different teams? Is technology working to reduce the risk of operations years down the line? Are there any bloated or understaffed teams?

Ideally, the technology leader should clearly demonstrate to the CFO (Chief Financial Officer) and CEO (Chief Executive Officer) that, for every dollar invested in the company's technology team, there is a greater return on the initial investment. They could achieve this, for example, by analyzing the difference between the cost of maintaining and operating an in-house system and a viable alternative provided by a third party. Sometimes, the difference amounts to millions of dollars a year, especially with scale.

In addition, any improvement that a technology team makes to the systems of a company that is scaling up its processes, no matter how small, can generate real value for the business. For areas that are strategic for the company, developing the technology in-house also generates advantages, such as greater control over the tools, customization of the technology to the company's needs, and the possibility of innovation.

A certain amount of "fat" in a technology team is desirable so there is room to fail, try again, experiment, and find innovative solutions. Innovation should always be at the heart of any technology company. The only people who never make mistakes are those who only work on obvious projects. Mistakes are part and parcel of everyday life in technology. In some projects

I've worked on, such as search engine efficiency,[8] it was necessary to carry out many experiments for each incremental gain. However, a gain of even 0.5% in efficiency generally represented millions of dollars, given the scale of these systems.

If there is no money for this "fat" at an early stage in the company, it can be created by efficiency gains in the technology area—for example, by making developers more productive, reducing technical debt, and reinvesting this productivity gain. But you also must pay attention to how quickly the technology team grows because the bigger it gets, the more complex it will be to manage.

While the team is smaller, it's easier to manage what each person does and to be more judicious when recruiting. You can no longer have individual exchanges with all employees when you get to larger numbers, from 50 or 100 employees and up. Anderson Nielson, who works with me at Stone as my HR business partner, always says that there comes a time when the company becomes a reflection of society, where there is everything, including less efficient and more efficient people, and people who are less adapted or more adapted to the culture of the technology and of

---

8. I worked in search engine efficiency during several years and in multiple companies (IBM, Yahoo, Google, and Microsoft). At the time, even while sleeping I had dreams about how to optimize inverted indexes, which are the fundamental data structure for search engines. They were very interesting dreams, but they often didn't work out.

the company. At this point, there is a need to establish control and supervision processes, standardize platforms and tools, and define clear guidelines. Orderly and well-thought-out growth avoids the painful operation of taking away freedoms and imposing restrictions on a team that is already used to doing everything its way.

A **platform mindset** is thinking about all these goals, with intentionality: extracting the most value of the technology team while maintaining a great culture, platforms, and tools, and reducing technical debt and inefficiencies. Of course, this must be done with a growth mindset, as there are no simple formulas, and we need to learn as we go. The following chapters provide insights into many of these areas, based on my experience working in innovative teams and leading large organizations.

Changing the game

The technology leader isn't necessarily an insightful person who can invent a new application and knows all the latest tech news on the market. To lead a technology team, you need to be someone who can talk to colleagues from all areas of the company to understand how to make technology an integral part of each area's function. Many people think that this means creating systems to automate manual processes. It is more than that. Digitizing manual processes is like wasting time thinking about making a wagon faster instead of concentrating on building a car.

Technology needs to **change** the game, not just speed it up. If each department in the company hires a few technology professionals to separately create ten systems, you'll end up with ten problems. Each of these departments may be better locally than before, but the company overall will be in a situation far below its potential due to a lack of communication and integration.

The case of the distribution of COVID-19 vaccines in the United States during the pandemic proves how not thinking intentionally about innovation hinders project execution. Once the long-awaited COVID-19 immunizations were manufactured and made available to the public in early 2021, the US government distributed them by sending batches of vaccines to the country's pharmacy chains to manage their administration. These large private chains then divided the shipments they received among their branches. As a result, some areas of the same city had more vaccines and others less, possibly depending on the number of different pharmacies and residents there. For the consumer, it was necessary to go to the website of each pharmacy chain to search and find out in which region near their home or work they could schedule their vaccination.

If a single distribution and information system had been created, the service would have been much simpler and more convenient for the user/citizen, as well as more effective in reaching the government's goal of vaccinating as many people as possible in the shortest possible time. A single central system could speed up the population's vaccination schedule. A detail like that makes an absurd difference to the number of vaccines administered and the streamlining of the whole process.

As the scale of the COVID-19 pandemic was unparalleled, no one had planned a vaccine distribution protocol that could be quickly implemented and scaled up. The process wasn't designed to make technology into an active and essential part of what was being done, and it's precisely in this kind of situation that it's possible to see how a lack of strategic technical direction impacts the desired results.

Technology can't be an accessory to the project; it must be central to it. There's no point in developing a service or product and then thinking about how technology will fit into it. With a technology person involved from the outset, the possibilities for execution are multiplied. It's not about improving an originally analog process but transforming it.

There are situations in which the company's resistance to the implementation of this technology leadership model is so strong that, even in the face of a well-intentioned and informed manager, the digital transformation stalls. Often, there is no room in the company for a technology leader to head an ambitious project because no one there really understands the role of technology in business. A platform mindset may require a cultural transformation.

Changing the culture

Microsoft was born in 1975, before the internet, when software was sold in boxes, and there was no concept of online services or cloud computing. The company is still relevant today, and perhaps more than ever, because it is constantly reinventing itself. CEO Satya Nadella's insight was to understand that their old model of thinking, in which each part of the company ran its projects as if it were an independent business, even competing internally with other departments, no longer worked. Satya realized it was necessary to generate synergy to maintain a leading role in a new environment where everything is connected.

I worked with Satya at Microsoft for eight years, and I saw how dedicated he was to changing the company's internal structures. He pointed out to the teams that they should collaborate and depend on each other, such as betting on Azure as the common computing platform. This brought the people in leadership into a dialog that hadn't existed before and generated unprecedented synergies and experiences for everyone.

As part of this transformation, I worked with Mark Russinovich, CTO of Azure, on the migration of internal services that used their own computing platforms to Azure. These conversations were very difficult, almost non-existent at the beginning of our journey. However, throughout the company's cultural transformation, they became more frequent to the point of bringing unquestionable value. The evolution of Azure as a product helped the viability of these migrations.

Still, they were only made possible because the various teams at Microsoft began to collaborate more and became comfortable with being more interdependent.

In his book *Hit Refresh*,[9] Satya describes Microsoft's cultural transformation. However, each organization presents its challenges.

The role of the technology leader is to envision, communicate, and implement the mechanism that will facilitate change, gaining the cooperation of employees along the way by showing that the future holds bigger incentives. This involves working with the technology team and the rest of the company, which needs to start seeing technology as a partner area fundamental to transforming and scaling the business.

There are many books on organizational culture, and that's different from the focus of this book. However, in the context of the technology leader's mindset, I'll talk about various aspects of culture later in the book, including how to attract and keep the best people for your team, how to organize the team, and engineering processes and tools. All these things impact the company's culture, and maintaining culture is a daily exercise. As Satya says, the day we declare the cultural transformation over, it will start again—the company and the technology organization are living organisms that require constant care.

---

9. *Hit Refresh: The Quest to Rediscover Microsoft's Soul and Imagine a Better Future for Everyone*, by Satya Nadella (Harper Business, 2017).

When I joined Yahoo, I was one of the first employees of the research department built and led by Prabhakar Raghavan, currently senior vice president of engineering at Google. Prabhakar said that our job as a research team was to transform Yahoo's culture into one in which scientific thinking prevailed. In practice, this involved not only growing our team from the ground up to encompass hundreds of researchers in a few years but also the way we interacted with the rest of the company.

One of the earliest projects that helped us establish the importance of thinking scientifically about projects happened a few weeks after I joined the company. The search team faced the issue of "index size"—at that time, search engines advertised how many web pages they had in their index, and the competition was fierce amongst Yahoo, Google, and everyone else. The outside world didn't necessarily believe in the numbers advertised by the search engines and wanted to estimate their index size independently. We worked on the problem and developed a rigorous method that enabled outside teams to do this evaluation and later published the algorithms and the method in a scientific paper.[10] With this success story and others that followed, after a while, the leaders of the other areas

---

10. Estimating corpus size via queries, A. Broder, M. Fontoura, V. Josifovski, R. Kumar, R. Motwani, S. Nabar, R. Panigrahy, A. Tomkins, Y. Xu, *Fifteenth Conference on Information Knowledge Management* (CIKM 2006), 594–603, Arlington, USA, 2006.

wanted to involve our department in their projects, so we gained ground and became crucial for the company. There was clear value being created. The company became a place where scientific thinking prevailed, not because of a decree by the CEO but because of the quality of the systems we had built and how they integrated with the rest of the company's systems. One project at a time, in a gradual process that ultimately had a huge impact: all of Yahoo's ads were now being served by systems developed in research. Not to mention ZooKeeper, an open-source system still widely used today, also developed there. In other words, their bet on people working more exploratory and with freedom to create ended up paying off in several areas, including products and services that we didn't even know would become very successful later.

IMPORTANT POINTS REGARDING
PLATFORM MINDSET

- Promote an integrated vision of technology for the organization, rethinking the main processes, especially where there is an intention to scale up.
- Balance between taking care of technical debt and developing new functionalities and systems in a conscious way.
- Develop a well-defined strategy on whether to develop in-house or outsource.
- Leave space for the team to experiment and innovate.
- Encourage disruptive changes, which are riskier but can make sense.
- Be aware that technological changes may require a cultural transformation.

Chapter 2
Focus on people— the "artists"

Throughout my career, I have witnessed the focus that big techs place on recruiting the best talent for each area and experienced in real life how this strategy boosts the business. In companies like IBM, Yahoo, Google, and Microsoft, executives in the highest positions are personally involved in attracting and hiring the best people.

In addition to hiring, developing people's abilities and managing their careers in a structured way, it is essential to enable the engineering team to function as a high-performance team. Having the right people assigned to projects that motivate them and align with their skills creates tremendous value for the company. I often joke that a team of exceptional people, whether hired from outside or developed in-house, makes the engineering leader's job very easy. The problem is that a team of exceptional people is difficult to build and maintain.

In this chapter, I'll cover hiring, mentoring, evaluating, and managing people. An important point is the Y-career path, where individual contributors play a crucial leadership role. I'll also discuss the importance of inspiring people and managing their careers with a long-term perspective.

Hiring the best

The presence of inspiring professionals in a team has a positive impact and hiring them is a good investment. First, such a person will likely design and implement a brilliant and innovative idea that revolutionizes the business, brings huge savings, and gives rise to a new service or branch of activity. These events are rare, but they do happen, and organizations need to make room for opportunities of this kind to arise.

Beyond this potential, inspiring people act as aspirational role models who help establish the company's culture.[11] As mentors, they work to form new leaders within their teams and act as magnets, attracting other good employees, either because they are followed wherever they go or because they have good criteria for selecting future hires.

Candido Portinari is considered one of the most important Brazilian painters and a prominent practitioner of the neo-realism style in painting. When, in 1952, he was commissioned by the Brazilian government to paint the "War and Peace" panels that were to be gifted to the United Nations (UN) to be exhib-

---

11. In his book, *What You Do Is Who You Are: How to Create Your Business Culture* (Harper Business, 2019), Ben Horowitz describes the role of role models in setting the organizational culture.

ited at their General Assembly headquarters in New York, no one said to him: "Paint red over there and black over here, make figures in such and such a way in this or that quadrant." He was the artist, and whoever commissioned the work trusted his talent.[12]

I think of engineering professionals as artists capable of practicing their art in the areas where they work if they receive the freedom and conditions to do so. It's not enough to hire good people if they don't have a certain level of independence and the environment to perform at their best. Everything flows naturally when professionals are motivated and in tune with the company's mission.

It's the technology leader's role to make this cycle of talent and execution happen. The problem is that fostering the ideal work environment takes effort and, more than anything, money. However, I believe it's worth investing as much as the company can in its top talent.

In engineering, there is the idea of the "10X professional"—a person so efficient that they end up

---

12. The letter, the panels, and the studies are available on the Portinari Project: https://artsandculture.google.com/story/dAVh9lXSqG6hJA?hl=pt-br. I had the honor of working on the beginnings of this project, which among other initiatives digitized Portinari's collection, during my undergraduate studies. João Candido Portinari, the painter's only son, is a mathematician and one of the founders of the Mathematics Department at the Pontifical Catholic University of Rio de Janeiro (PUC-Rio), Brazil.

doing work equivalent to that of ten average employees. When this happens, that employee, even if an expensive resource, leads to savings. And the effect keeps replicating itself because super-efficient employees attract other super-efficient employees, while mediocre employees bring in mediocre employees.

Many startups already begin with a few super talents among their founders. It's unusual, however, to be able to develop such talent within the company itself. For this reason, the fight for talent in Silicon Valley, for example, is enormous. Everyone looks for outstanding people, those with high achievements on their resumes because they have already been tried and tested and will continue to produce a huge impact on the companies that are lucky enough to hire them.

Multipliers and diminishers

According to Liz Wiseman and Greg McKeown's concept[13] multipliers are those who provide the conditions for their team to do their best work and who let their employees learn from their mistakes. They stay close only enough to avoid disasters without intervening at every stage.

Multipliers see things from the company's perspective and usually have a collaborative mindset without acting in an exclusionary way: they involve other areas

---

13. *Multipliers: How the Best Leaders Make Everyone Smarter*, by Liz Wiseman and Greg McKeown (Harper Business, 2010).

of the company, create efficient and sustainable processes, and bring in new talent. They end up getting the best out of the employees under their influence. They are the opposite of the so-called diminishers, characterized by their micro-management and focus on control and self-promotion. The technology market has a lot of individualistic people who got used to being treated as the most intelligent person in the room during their education and then turned into centralizing and controlling professionals. I've lost count of how many prima donnas with these traits I've met in the places where I've been.

Depending on the situation and career stage, we can all act as either diminishers or multipliers. It's a daily exercise to understand when we are acting in one way or another, the reasons for our behavior and what we can do to change course and be more effective.

Diverse talents

One way to attract good professionals is to build a reputation for providing an environment focused on learning and challenging day-to-day activities. This certainly happened with Google when it became famous for having a group of incredible people developing the best search engine in the world. Everyone wanted to be a part of that.

I can include even myself among those who saw Google as an opportunity to learn like never before. I had been offered a tempting opportunity to work at a promising startup. Still between the chance to learn new things at Google, interacting with experienced professionals, and the prospect of being one of the most senior employees at a startup, where I would have fewer chances to learn and grow, I decided to join Google. It was a decision that may have weighed heavily on my wallet at the time, but it paid off over time with the unique experience I had, working in that effervescent environment.

I know that there is only one Google and, if it were easy to do what they did, their business model would have been maintained indefinitely and spawned many other companies just like it or better. However, it is possible to foster a corporate culture that values professional excellence, even if on a much smaller scale.

Working in an exciting environment with great colleagues can be attractive, regardless of remuneration. I've seen people come to Stone to earn less than they used to, and I'm excited about the possibility of helping build the company and get it to achieve its mission of building technology to transform small and medium-sized businesses.

From the manager's perspective, it's more complex to identify the right people in the market. Some attributes are easier to measure, such as technical background, which is clear from the resume and achievements. On the other hand, it's essential to detect people with personality traits that make teamwork difficult, for example. It is plausible that few specific projects could benefit from non-collaborative hires, but the leader must be aware that they will not be an inspiration to other employees.

In the companies I've worked for, I've met people who were equally good in terms of technical quality but with opposite behaviors: some of them stayed locked in their offices, working on a significant individual breakthrough, while others liked to collaborate and share their knowledge. For a balanced staff in the company, we're looking for a combination of qualities and characteristics, not just in individuals but in teams.

When hiring for a senior position, it's the most natural thing to resort to the personal networking of the people in leadership roles. The risk with this approach, however, is that you end up hiring professionals very similar to those already with the company: people with the same educational background, who worked in the same companies, with the same skin color, and have the same way of thinking. This removes the element of diversity, which fosters new ways of interacting, producing, and creating competitive advantage. The possibility of new insights drops a lot in non-diverse environments, and nowadays, this type of company, where everyone looks alike, is no longer attractive to an increasing number of professionals.

The United States' history of immigration has contributed to a highly qualified and diverse workforce in the country's technology companies, which have had the luxury of hiring people from anywhere in the world to join their ranks.

The value of interviews

What you can do to avoid falling into the trap of hiring "the same" type people is to interview as many people as possible and not be in a hurry to fill a vacancy straight away. When I'm hiring someone, I think in the long term. Sometimes, the person I've chosen needs time to get organized before they can move to another company. At Microsoft, I used to say to successful candidates: "We've waited 40 years to hire you. We can wait another three months." Filling a senior position is an investment of not only money but also time and dedication from the most influential people in the company, which is why I advocate that it should be done with patience and care.

"What's your relationship with your team like?" and "Do you think people will want to follow you when you come here?" are some of the questions I often ask in interviews to assess candidates' ability to inspire.

Interviewers should remember that the person they are talking to may not have succeeded in their previous job because there wasn't a match between the professional, the role, and the company. Even so, depending on the candidate's personality, skills, and even personal tastes, you can see that perhaps that professional can become someone with whom younger employees will feel inspired to work with.

Nowadays, it's also crucial to avoid getting carried away by a public impression based on social media profiles. Some people seem inspiring because they have a lot of followers on social networks, with profiles that give good explanations on technical subjects, but that doesn't mean that this ability will play out the same way within a company. There's a risk that they won't be good at working directly with people or they'll be more concerned with making a name in the digital world than dedicating themselves to the company.

I check references from previous jobs and try to "sense" a bit of the person's charisma in the interview. For more senior positions, the interview is longer because more people talk to the professional. Google and Yahoo interviews were famous for focusing on reasoning questions and determining if the candidate was an independent thinker. It's as if they were trying to find out the power of the person's CPU and their processing capacity. However, understanding other aspects of their personality is more complicated than that.

I like to be guided by my intuition about a person as soon as they walk in the room, by the way they present themselves, but that's not all I analyze. The candidate needs to be smart, insightful, interesting, and interested. They need to be aligned, or at least have the potential and the desire to be aligned, with the team's collaborative spirit. It is essential to assess whether they will commit themselves in the long term to the company's philosophy and work ethic. This is the most challenging part to assess in an interview environment because it's a very short time to look for any warning signs or slip-ups.

At Yahoo, I once had a lunch interview slot for a candidate. It was common for the interview to last all day, and one of the interviewers had to accompany the candidate for lunch. To my astonishment and horror, he was vulgar to the attendant in the company cafeteria, who didn't understand his lunch order correctly. That made me wonder how easy it would be to collaborate with him. After he got his terrible lunch (mine was pretty good), we started talking about work and life, and I noticed his arrogance. The moment I saw that, it was an automatic rejection from me. Several of the other interviewers sensed the same behavior during their interviews. We unanimously rejected the candidate, even though he was technically brilliant.

While it is possible to see traces of arrogance in some situations, you should bear in mind that, in technology, it is not uncommon for people with great potential to be eccentric and neurodivergent. In these situations, interviewers must be sensitive to whether they can play the company's game and whether building a team around them or with them will be possible.

At Google, we treated the interview process like an algorithm, with eight evaluators who independently formed an opinion about the candidate and then came together to exchange assessments. If one of us had any restrictions on hiring, we would err on the side of caution and reject the candidate.

From the point of view of the person being courted, the personal commitment of the company's leadership makes a big difference. In 2011, when Microsoft took over Yahoo's search, I decided to make a career transition. I interviewed at many places, including Google, and then I received an invitation to dinner from Alfred Spector, the company's vice president of research.

He didn't have to take me to a fancy restaurant to make a deep impression on me. I saw in him, at that dinner, what I would take as an example for myself: a leader of renowned technical ability and involved in practical projects who showed me that he acted strategically. He was a role model if I wanted to become a leader in systems research, but at the same time, someone who was very attentive and took the time to study my resume carefully, to the point of recognizing the gap that existed in my scientific production around 2006.

He asked very nicely what had happened that year, why I hadn't been able to produce as I had in previous years and was extremely polite when I told him the reason. At the time, my wife, Renata, had undergone seven surgeries and several aggressive chemotherapy treatments to fight against breast cancer, which would eventually prove fatal. His attitude and respect towards me were decisive in my decision to join Google.

Selection process and diversity

There are often different selection processes for senior positions and the pyramid's base. For the most senior positions, the responsibility generally lies with the technology leadership, who knows the market's "artists."

For entry-level positions, the recruitment team is usually responsible for attracting new talent. This team usually relies on mass selection initiatives, such as university recruiting, trainee programs, and internships. This type of hiring mechanism offers the opportunity to have a more significant social impact, allowing companies to attract young talent from underprivileged communities.

Also, to encourage a diversity of thought, we at Stone have adopted initiatives to sponsor the training of people from other fields who want to work in technology. In one of these projects, we selected 200 people from professions such as law and architecture and funded a six-month programming course for them. At the end of the training, we divided the class into teams that had to implement projects conceptualized by Stone employees, who also worked as mentors. We held an event on a Saturday with all the company's leadership present to evaluate these projects and hired several of the sixteen finalists. The ones we didn't hire were in a good position to interview for development positions in other companies.

Initiatives like this help create a strong employer brand that attracts entry-level and mid-career candidates. The company's name must be publicly associated with an environment of education and career development. This conscious effort to solidify the employer brand includes sponsoring events that match the company's philosophy and getting closer to the academic environment with established internship programs and academic partnerships.

It's common in hiring processes for candidates to be asked programming challenges or brain teasers like

"How many marbles can fit in an Olympic-size swimming pool?" or challenges with open-ended questions about architecture, such as "Imagine you need to make a system to store the personal data of all the people in the world. How would you do it?"

With these questions, we need to look for the right answer. The intent is to allow interviewers to see how the person starts to think about the problem and whether they can reason about it—a bit of what I mentioned earlier, trying to assess "processing capacity" (the CPU size). We also give very concrete programming tasks to see if they are independent in the basics of programming, to align with how much we are willing to invest in basic training for specific positions at that moment, according to our needs.

Choosing young, inexperienced people to work for the company is more or less a shot in the dark. Since that's the case, I'm in favor of adopting the bias of the potential social impact we can have, placing our bets on those who come from the neediest areas. A salary from a technology company, even at entry-level, in an underdeveloped region can revolutionize the standard of living of an entire family.

Initial support

In the early 2000s, it was essential to have a degree from an American university to enter the competitive job market in the United States. I was disadvantaged because all my training had been done in Brazil, more precisely at the Pontifical Catholic University of Rio de Janeiro (PUC-Rio). It's an excellent institution that gave me a solid education. Still, those who take undergraduate and postgraduate courses at US universities are naturally closer to American companies and can more easily participate in their internship programs. I didn't have this experience, and I didn't nurture any professional network during my studies.

I took the PUC-Rio entrance exams in 1991 to study mathematics after falling in love with the world of numbers in the final years of elementary school. It wasn't until I took my first computer science course that I found my real passion and changed my emphasis to computer engineering. It was an eventful time because, in my first year at university, my father was diagnosed with lymphoma. Since he was ill and undergoing treatment, I didn't consider moving away from Brazil for my master's and doctorate. The alternative I found, together with my advisor, Professor Carlos José Pereira de Lucena, was to pursue my graduate studies at PUC-Rio while following the developments of my father's illness.

The chemotherapy treatments went on and on for several years. At a certain point, the doctors announced to the family that they would try one last new medication that had become available. Either it would work, or there would be nothing more that could be done. The medicine worked—my father is healthy to this day! With his cancer cured and a stable situation on my family front, I left for a one-year post-doc at Princeton, in the United States.

Even though I attended the prestigious Princeton University during that period, my Brazilian-based education didn't excite the companies and universities I applied to, given the lack of responses to my applications. So, I decided to email Jean-Paul Jacob, a Brazilian engineer who held an important position at IBM Research and with whom I had never spoken in person.

With a brilliant career at IBM's Almaden Research Lab in California, Jean-Paul was generous enough to reply to my message. Through him, I got an interview and was hired to work at IBM, where I began my career in the United States.

Like Professor Lucena, Jean-Paul would become a great inspiration to me and a definitive influence on my professional development. Jean Paul's specialty was interfacing between research and the general public. He was a born communicator, a very entertaining person who could explain to non-techies how computers and the internet would revolutionize the future.

I always try to replicate Jean Paul's attitude toward me: talking to everyone who approaches me, giving opportunities, and being accessible. I also set an example for the professionals I hire and mentor so they can cultivate this open attitude with their reports and mentees.

Mentoring and development

The effort to hire curious, committed, inspired, and inspiring professionals is crucial, but it is only the first step. Once people with potential arrive at the company, no matter the level of their position, we need to keep them interested and productive. This will ensure that they stay with the company in the long run and that the investment in hiring is not thrown away with an early departure. In a market like technology, employee mobility is high, and the lure of other companies could even be aggressive in their attempts to attract the best talent.

Integrating a new employee into the company, known as onboarding, is crucial. It is the new person's gateway into the company and when they form their first impressions. Several studies show that this initial period of employment determines whether the person stays, as up to 20% of new hires leave within the first 45 days.[14]

The most well-known part of the onboarding process is also the most obvious: introducing the new person to everything, from the systems in use to the company culture, tools, and processes. Organizations have been improving at creating a good onboarding experience. Still, it's worth stressing that there's much more at stake than giving a lovely welcome gift, scheduling one lecture after another, and making the person feel welcome. Of course, this is necessary, especially

---

14. "To Retain New Hires, Spend More Onboarding Them," by Ron Carucci, in *Harvard Business Review* of Dec. 3, 2018, available at https://hbr.org/2018/12/to-retain-new-hires-spend-more-time-onboarding-them (accessed Mar. 28, 2024).

when the work system is remote, and there is a physical distance to overcome. I like to get personally involved in this first week by using the Ask Me Anything (AMA) chat I have with all the newcomers at Stone.

In this chat, I tell newcomers about my career, discuss my vision for Stone, and allow them to ask me questions about anything—even soccer, in which case I'm not able to answer them well. This first contact is crucial for creating a bond and an initial relationship with everyone joining the team.

However, the onboarding process in a technology company will only be complete when the incoming team members start to feel productive, regardless of their position. For this to happen, in addition to the onboarding process, ideally, the new hire should have a person dedicated to mentoring them, and they should be immediately involved in an initial project. This project, often known as a *starter project*, should last a few weeks and be comprehensive enough to expose them to the development methodology, the code review process, and the central systems, tools, and libraries they will be working with daily.

In this initial phase, the new hire's team will pay close attention to the code they write. The code reviews will be done more assertively to teach them the company's best practices. In the engineering lifecycle, employees write code, test it, and submit it and its review test, allowing coworkers to comment on it in an iterative process until the code is good enough and can go into production. This process of validating and reviewing code is a learning process, especially for newcomers.

When I joined Google, everyone on the team commented on my first code reviews because there was a robust culture of code zeal there. It was tough to get an "LGTM" (Looks Good To Me), which indicated the end of the review when everyone was satisfied with the code quality. As in other large technology companies, code review is an opportunity to teach new people the processes and how the system is organized in detail. It's a significant process for newcomers.

In smaller companies, the code review culture may be less widespread, unlike in the big techs, where it is extensive. Some people, especially senior professionals, think educating people who came into the company after them is separate from their duties. Perhaps they don't know that at Google, for example, at least while I was there, leaders spent around 60% of their working time reviewing code, and all developers were copied on every review. Reading the detailed comments in code reviews, both mine and those of my colleagues, made me a better developer. For the leaders, it was a fundamental part of their job because, besides functioning as a training ground for the workforce, this is a way of seeing areas that require special care, teaching people how to write tests, and, in short, determining what the architecture of the system will be. Through code, the leaders technically communicate what they want to be done in the organization.

You can't look at code review as a bureaucratic task: someone else has submitted some code, and you need to approve it so that it goes to production without reviewing it thoroughly, without any comment, and without any criteria. When you use code reviews as a teaching opportunity, the team's skill level rises and the bar is raised. In the future, these same people will be capable of conducting constructive code reviews. It's a way of leveling up the team and an aspect of the engineering culture that I've worked hard to bring to Stone.

Transfers in senior positions

Even a senior person can feel lost in a new company's early days. One of the main factors that makes this integration difficult, for instance, is the team's vocabulary—something basic but profoundly affecting day-to-day life. The new person needs to learn the team's terminology, the names of the systems, and the acronyms.

Managers are responsible for ensuring that new hires are mentored and supported until they feel comfortable in their new role. Weekly one-on-one meetings are a good tool for this. Hiring a talent coveted by the market is a considerable investment and a victory for the company, and it can all end up being wasted if the newcomer doesn't adapt well to their new role.

One of the things that generates the most anxiety is when the professional can't see where they stand concerning expectations. Leadership needs to be accessible and transparent, as transparency helps to reduce insecurity. All communication opportunities should be used, from internal messages to all-hands meetings. Regular and constant communication can be adapted to the company's management style. Even if the management is much more bottom-up than top-down, frequent communication sets the tone for technical and strategic guidelines. "I'm on board, but I want to know where this ship is heading"—that's a very fair consideration for those who have joined the company.

It's very uncomfortable not to understand how the company works, and the managers need to reassure the person that it's okay to take some time to feel at home. I experienced this first-hand with a job transfer I went through. It wasn't really a change to a new company because I was already at Microsoft, but I had spent years and years working on search engines, and suddenly I transferred from Bing to Azure, the company's cloud computing area. The shake-up was deliberate. The head of Azure Compute, Girish Bablani, currently a president of engineering at Microsoft, wanted to bring in someone with a different mindset. Many of the people who worked there came from Windows and had an operating systems mentality, while my background was in search engines and distributed systems. His idea was to throw me into the fire, and I agreed.

As chief architect for Azure Compute, I was responsible for establishing direction in the area. But first, I needed to learn how things worked. My strategy was to attend meetings of one of the central teams, as if I were one of its developers, and try to understand what they were doing. The first few meetings I attended, I left without understanding anything. My plan worked out well only because I had the support of Girish and my peers. And because I worked on concrete projects while I was learning, even if they were simple projects, enabling me to deliver some results early on in my new role. No one who is a good professional can go for a long time without contributing in some way. This is similar to giving a starter project to someone who has just joined the company. Employees are happier when they have the space and conditions to feel useful, productive, and valued. This creates a **positive cycle** because when someone feels happy at work, they are willing to create more, mentor other employees, and be a role model. Over time I learned a lot about my new area, Azure Compute, and contributed new ideas that have had a high impact, increasingly leading larger and more complex projects.

Y-career path

A widespread mistake that technology managers make with young employees who are starting to excel at work is to promote them too early to management positions. The idea that being promoted means assuming an administrative role is widespread, with the mistaken view that prestige, higher salaries, and impact are closely linked to the number of people you manage.

In technology, putting someone into a manager role too soon is a terrible thing to do to a promising person. One of the first steps we took at Stone after my arrival was formalizing the so-called Y-career path.[15] It has a joint base for all engineers. Then it bifurcates into one arm for traditional management positions and another arm for individual contributors, also called specialists or Staff+ engineers, as shown in Figure 2. Individual contributors move up the career ladder parallel to managers, with equivalent pay and prestige. The difference is that their role remains more technical than managerial. Individual contributors (ICs) are responsible for giving technical direction to their areas, mentoring more junior employees, establishing the engineering culture, and writing and reviewing code, often working within the scope of a specific project but without having the administrative obligations of performance reviews or approving airline tickets.

Hiring, developing, and celebrating individual contributors is one of the best ways of highlighting the company's technical talents. The challenge is to legitimize the individual contributor position to the point of making it attractive to outstanding professionals.

---

15. To spread the culture of a career plan that includes individual contributors in Brazil, we made a point of making public the *Star Trail*, the career trail we created for Stone. It resembles, and is inspired by, the career trail of many Silicon Valley companies.

Figure 2: Example of a Y-career path in technology

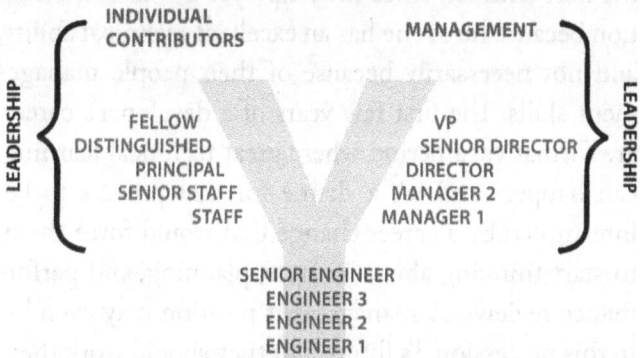

INDIVIDUAL CONTRIBUTORS	MANAGEMENT
FELLOW	VP
DISTINGUISHED	SENIOR DIRECTOR
PRINCIPAL	DIRECTOR
SENIOR STAFF	MANAGER 2
STAFF	MANAGER 1

LEADERSHIP

SENIOR ENGINEER
ENGINEER 3
ENGINEER 2
ENGINEER 1

I'll make a controversial statement here: in a technology company, management matters, but it's not what leads to real differentiation. Management organizes the game and allows wins to happen, but its main function is not getting in the players' way. I've spent most of my career as an individual contributor, and today I'm in management as a technology leader because I wanted to work more directly in changes to corporate culture. In Stone's case, my mission is to implement the platform mindset of a big tech in a Brazilian company, which thinks intentionally about technology and values individual contributors.

Normally, people in technology start their careers as engineers on the base of the Y. Nobody becomes a manager as soon as they leave college. Once a professional starts standing out by becoming an expert developer, there's a risk that they'll soon be put in charge of managing people. However, this is often not the best decision since the employee has drawn attention because he or she has an excellent technical ability, and not necessarily because of their people management skills. The first few years of a developer's career are formative, a period where great technical learning can happen, and it's a shame for that process to be interrupted by a career change that would force them to start thinking about budgets, planning, and performance reviews. A management position may even be in this professional's future, but they should work their way up through the ranks, in the base of the Y, until they reach a more mature point in their career. This will bring better results for the company and the person's development.

Another area for concern is that some management skills are not easily transferable to another company. A young, promising employee learns, for example, how to work with the human resources system of their company, and, when they transfer to another organization, they may find a very different type of performance evaluation system. Since they lack experience, they cannot even have a more strategic opinion on how performance reviews should be carried out. This early interruption in technical learning can harm their career in the long term.

When the company is tiny and still has a startup mentality, people don't care what position they hold and end up doing a bit of everything. They are embarking on something new, and no one worries about titles.

This motivation compensates for the disorganization and lack of a career plan. However, once the company passes a growth threshold, people want to know where they stand. In a large, organized company, the employees should have a career plan from the start, with defined duties.

The role of individual contributors is, by its nature, ambiguous. It's great when companies formalize the Y-career path. Still, even so, the positions are wide-ranging, with loosely defined functions that vary greatly depending on the professional's interests and abilities and on the company's organization. The person can act almost as a manager, leading teams and mentoring people—without administrative duties—or they can act as a senior developer, writing code for a more complex part of the system. Or they could operate between these two extremes. There is a risk that they may feel isolated or lost, even though they are experienced and part of the company's technical leadership. One of the initiatives I started at Microsoft to alleviate this problem was to create a community of senior individual contributors in the Azure organization, with active communication, lectures, and a structured program to facilitate dialogue and the exchange of ideas between them.

When you're a senior individual contributor, all you want to do is solve a relevant problem. The community we created at Microsoft allowed its members to collaborate and exchange ideas, which made their work a lot more manageable.

It's worth pointing out that employees can change from one branch of the Y to another. It's good that individual contributors may decide to spend time on the other side, as managers, and vice versa, in changes that could become definitive if they feel so inclined. I'm an example of someone who has made this transition, going from individual contributor to manager, and there are professionals with a restless, challenge-driven spirit who would benefit from making the pendulum swing between the two career paths, spending periods on either side. There's nothing wrong with that.

For those more interested in this topic, I recommend two very relevant books on developing people in technology. They are somewhat mirrored works for each arm of the Y career path. In 2017, Camille Fournier, vice-president of technology at Goldman Sachs, wrote the book *The Manager's Path: A Guide for Tech Leaders Navigating Growth and Change*.[16] In response, or as a complement, engineer Tanya Reilly, with more than 20 years of experience as an individual contributor at companies such as Google and Squarespace, has released *The Staff Engineer's Path: A Guide for Individual Contributors Navigating Growth and*

---

16. *The Manager's Path: A Guide for Tech Leaders Navigating Growth and Change*, by Camille Fournier (O'Reilly Media, 2017).

Change,[17] aimed at those interested in traversing the technical arm of the Y. These two books give a detailed overview of the career possibilities in technology, from the standpoint of the engineering professional pursuing these different careers, manager and individual contributor.

Inspiration and superpowers

The key people in an organization, those I referred to at the beginning of the chapter as the "artists," serve as a beacon for the professional aspirations of all employees. They act as inspiring role models, not only for retaining those who are already there but also for attracting new hires. When there isn't a leading figure in the arm of individual contributors, there may be fewer people looking for this career path, and the technical quality of the organization may suffer.

The people in the spotlight tend to give the company a specific identity, and it is the managers' responsibility to give them this visibility. The top management positions are prominent, but they are scarce: there is only room for one CTO. The top of the pyramid is narrow. If everyone wanted to be a manager because only those people serve as inspiration, there would be a limit to how much people could grow in their careers. You need key people, role mod-

---

17. *The Staff Engineer's Path: A Guide for Individual Contributors Navigating Growth and Change*, by Tanya Reilly (O'Reilly Media, 2022).

els, who belong to the technical arm of the Y. What can be done to sponsor and increase the visibility of individual contributors is, for example, to include presentations of their work in the company's monthly meeting and talk about their projects in communications with the team.

When one of Stone's technology managers asked to become an individual contributor, I told the whole team about the change in my weekly post to the technology organization and celebrated his decision. I intended to demonstrate the importance of his transition and emphasize the two parallel possibilities for career progression in technology at the company.

One of the key people at Google is Sanjay Ghemawat, who has been with the company since 1999 and today holds the position of senior fellow, the highest individual contributor title in the company. In my years at Google, I've never spoken to him directly, but that hasn't stopped me from feeling his positive influence in the company. Every time I looked at a piece of code, his name was there. People were inspired simply by the fact that he wrote so much of Google's technical infrastructure. As the code we were writing often affected the infrastructure, the code reviews and approvals automatically went to him, and he always gave his opinions very assertively. It was an expression of technical leadership and influence, exercised only via the keyboard. "I'd like to be the new Sanjay" was the standard thought that practically all my mentees at Google had. Sanjay had easy access to all the company leaders and was highly respected at every level—an exemplary figure.

The most inspiring example at Microsoft comes from the other side, from management. Everyone dreams of becoming the new Satya Nadella, the charismatic CEO who took over the company in 2014 and changed its direction.

It's human nature to be charmed by these extreme examples of success and to aspire to reach the same level. Sanjay and Satya are indeed exceptional professionals and have had the intelligence to direct their careers in such a way as to maximize their best attributes. Sanjay is a very talented developer. Satya has the gift of persuading people and is a visionary. Very few people will be as good a developer as Sanjay or as visionary an executive as Satya. Most people will fall somewhere in the middle. Mentoring people is about helping them understand where they are on this spectrum and how they can make better use of their abilities. The presence of inspirational figures like Satya and Sanjay serves as a driving force to show people that, just like both of them, they also have a superpower. Everyone has a superpower, even if it's not on the scale of a Satya or a Sanjay. The key is to identify each person's superpower and how to use it the best possible way. Promoting the **diversity of key people** within the company also helps people to identify with those whose profiles are more like their own.

In technology, there are many ways to contribute, because the roles are varied: maybe the person has the characteristic of being very organized, maybe they have vision and can establish new directions for the team, or maybe they are great at a specific type of algorithm, such as the ones used in machine learning systems. Some can have a panoramic view of the area and determine and dictate the architecture, which is my strength and the work I enjoy the most.

One of my sources of inspiration, albeit indirectly, was Scott Guthrie, one of Microsoft's executive vice presidents. I hardly interacted with him, but every time I attended a meeting where he was present, I felt like I was taking a masterclass on how to behave in meetings. He always came well prepared, having studied everything that would be discussed beforehand. He didn't blabber on and on, only making occasional but sharp contributions. When the team did well, he praised them and gave them the proper credit. When the job wasn't to his satisfaction, he gave constructive criticism. He was a true role model in management, an aspirational leader for me.

Skilled managers organize a positive environment with clear guidelines, expectations, incentives, and communication channels to enable the team to do its best work. So, when I think of artists within a company, I visualize not only the code wizards and individual contributors with their special projects but also managers and directors who masterfully run their teams' day-to-day operations.

Performance review

Managing professional expectations is a delicate process; the performance review system often ends up breeding bureaucracy and frustration. This occurs because companies need to evaluate what each of their employees does and what the impact of their work is, and the individual perceptions of performance are not always in line with this evaluation. For example, hours and hours of system maintenance work are undeniably important. Still, a creative solution that has sped up a process by 30%, even if it comes through a reasonably quick implementation, may be much more relevant to the business. This lack of direct correlation between hours worked and impact generated/perceived tends to poke egos and can lead to disappointments and the loss of talent.

Another major challenge of performance reviews is to make a fair and equal analysis for such radically different roles as senior management positions and individual contributors. In addition, as the roles and assignments of individual contributors can be fluid and undefined, it can be tough to compare their impacts.

I worked as an individual contributor for many years, giving direction to an area. When evaluating me for pay increases or career progression, my managers used metrics such as: "How many people are working on projects that Marcus has defined?" instead of analyzing how complex the algorithms I had developed were. Individual contributors can also work alone developing complex projects, such as distributed storage systems. In these cases, it wasn't possible to ask this same question—the question should be about the algorithms' complexity and impact.

With the Y career, a very senior individual contributor can be on the same level as a vice-president who runs an organization of several hundred people who, all together, may have a huge number of achievements. At the end of the year speeches, the vice-president may say he or she did it all, but it was mainly his team's work. The vice-president's achievement is different: having hired and managed great people, created an environment for innovation to take place, and invested wisely in short- and long-term projects.

The expert individual contributor, on the other hand, is the person whose solitary and sometimes time-consuming work makes huge advances or completely changes the game—and therefore requires quite specific impact assessment metrics.

To iron out some of the inconsistencies in the review process, I tried to establish at Microsoft, together with Girish Bablani, Alane Schettler, HR business partner, and Irada Sadykhova, senior director of organizational development, a segmented people review process. We first discussed all the individual contributors at a given level and drew our conclusions. Then we evaluated only the managers at that same level, bearing in mind that the impact of the work had to be measured differently according to their function. Talking about ICs and managers separately helped us more evenly compare people at the same level on each branch of the Y career path. A principal engineer may individually have more impact than an engineering director. Still, it's challenging to have that clarity when you've just discussed the promotion of a director who has described the results of his strong team of 200 people, which may have in itself some principal engineers and other senior talent.

Another point I was careful about was not evaluating and promoting people based only on how well they were known within the organization. I know a lot of great people who are extremely discreet or introverted yet manage to be inspiring and have an influence similar to that of Sanjay, the super-developer from Google.

Long-term people management

As a graduate student, someone told me, "Every doctorate student frets about their thesis, but the most important result of the Ph.D. program is not the thesis, but the student." I quote this phrase here to remind you that the professional, at any level of seniority, is what counts in the long run.

People management should be seen as a marathon, focused on the long term. However, one behavior that has become widespread is the adoption of very aggressive compensation packages, which fluctuate widely from year to year. If the employee hits their targets and has a good performance review in those twelve months, they receive a huge bonus. On the other hand, if they don't meet the requirements, they get nothing or are immediately fired.

It's like a 100-meter sprint for the company and the employee. This practice creates instability and is detrimental to retention, as highly rated employees search the market for the best chance of a bonus. In contrast, poorly rated employees leave the company, perhaps without having the chance to explore positions where they would be a better fit. Even brilliant people go through ups and downs in their professional lives.

On top of everything else, companies need to be flexible enough to deal with unavoidable personal factors that affect their employees. When my wife, Renata, was ill, my managers were aware of my situation, and my yearly review was not affected. I also received support from Yahoo after her death, granting me a sabbatical so that I could get back on my feet.

As a manager, I understand employees wanting to leave their current company to maximize short-term gains. However, this kind of move often interferes with learning in a limiting way. I've always strived to manage people's careers by thinking far ahead, and I have several examples of professionals who have stayed with me for many years.

This long-term perspective also applies when an employee thinks of leaving the company. When Andrei Broder, now a distinguished scientist at Google, was my manager at Yahoo, he had the philosophy of going as far as encouraging his employees to interview at other companies from time to time. For him, exposure to the job market had some advantages. One of the biggest was to help the employees understand whether they were happy where they were. A concrete transfer perspective eliminated possible idealizations: someone could dream of working at a new hot startup and, when they went for an interview there, realize that the environment wasn't what they had imagined or that the pay wasn't as high as they had heard. On the other hand, doing this type of interview opened new job possibilities, and the person could realize that the fit in their current company could have been better.

When a person decides to leave the company, I believe we need to help them do so. We need to look at it not only through the eyes of the organization, doing everything in our power to retain the person, but also through the eyes of the person themselves to understand what they are looking for in their personal development.

It is not benevolence that justifies this attitude of acceptance and understanding. Acting this way with those who leave the company strengthens the relationship with that professional, who will continue developing in other areas. They may end up going to a company with which we have joint projects or business relations or may even return to the organization. This person may also become a mentor to other talents and refer them to us in the future.

Ever since I left Microsoft, I've kept in touch and acted as a mentor to people who are still there. Of course, we don't talk about strategic or confidential projects, but we do exchange ideas on subjects such as possible career moves. It's a network that makes sense to cultivate, on the side of both mentors and mentees.

The habit of mentoring and being mentored is closely linked to the **culture of learning** and **growth mindset** that I advocate for. This kind of culture is intellectually stimulating and encourages the emergence of ideas and technical solutions that improve results.

However, a company's team of employees is made up of more than just high-profile role models. It's also a merit of people management to identify those who intend to advance to something other than leadership positions. They are professionals who work their way up from the bottom of the Y to the senior level and are happy to stay there. The point immediately before the Y's career crossroads needs a prestige and compensation level that is an acceptable final level for a portion of employees. The bottom of the career ladder is far more spacious than the top. The distribution of these roles and the organization of the team are crucial to the success of the technology team.

PRINCIPLES FOR ATTRACTING & DEVELOPING TALENT

- Organize the team around the best talents, giving them the space, freedom, and environment to be creative.
- Active search, both locally and globally, for diverse and inspiring talents, with a well-design interview system that allows for them to be identified.
- Promote an inclusive hiring and mentoring processes for the early career stages, avoiding the early promotion of developers into management positions.
- Define a Y-career path that values managers and individual contributors (ICs).
- Help employees identify and develop their superpowers.
- Stimulate work environment that is linked to innovation, to create a strong employer brand that attracts more talent.

Chapter 3

Team organization

I spent the first 15 years of my professional life working as an individual contributor since I first joined IBM. Neither at Yahoo nor Google nor for much of my time at Microsoft did I manage people properly speaking—I was leading very small teams at most. As I always enjoyed mentoring and leading technical projects and was inspired by the managers to whom I reported, my interest in management grew. I wanted to have the most significant positive transformational impact on the companies I worked for, and I knew that management would allow me to do that.

After just one year at Bing, Microsoft's search engine, Girish Bablani asked me to join Azure Compute as the chief architect, still as an individual contributor. I had no direct reports, and Girish saw me as a partner. While he managed the team, I took care of the technical architecture of the systems. Whenever we needed to restructure the team, Girish invited me to dinner, and we sketched out the new structure on a napkin.

Until the day he told me, at one of these dinners, "Marcus, the organization is growing a lot. Would you feel comfortable managing a team?" At the time, I held the title of distinguished engineer, a senior position in the engineering arm of the company. I said, "Yes," and that's how I, having managed a team of at most eight people, started managing a team of 400 people, which would soon grow to more than 1,000.

It worked out well because, although in theory I oversaw this entire contingent of employees, in practice I directly managed no more than ten people, my direct reports, who were the leaders of my organization's sub-teams. All of them were experienced and competent professionals to whom I could delegate much of the administrative tasks.

I started with this story to raise a basic concept of this chapter, which I consider fundamental for a technology company—and probably any company—organizational design. As part of this discussion, I will discuss management overhead and span of control, which is the number of people a manager directly oversees.

The more established the processes and protocols are, preferably in systems and tools, the less energy is spent on managing people. When the number of ambiguities is reduced, the team performance increases. In this chapter, I'll also discuss ways of organizing the team to encourage efficiency and creativity: the balance between more and less experienced professionals, the number of hierarchical layers, how to organize the team based on areas of activity and projects, and the importance of management rituals, such as the often-maligned meetings.

Management overhead

Imagine a technology manager with a team of 30 direct reports. The area has proliferated, with the company investing more in hiring young developers than managers. If this person wanted to closely monitor the team's work with individual one-on-one meetings of half an hour once a week with each person, they would have almost half of their work week committed to this task alone. If they then reduced the meetings to every two weeks, it is possible that, with cancellations and unforeseen events, they would spend more than a month without being able to exchange a few words with some members of their team.

Time spent on performance reviews would be another torment for them, with 30 evaluations to complete. They would spend so much time on management tasks, such as approving leave for each employee, analyzing possible promotions, resolving internal conflicts, thinking about retention packages, etc. They would be overwhelmed and hardly find time to do technical work.

What is happening in this example is that there is too much management overhead. Management overhead is the costs, mainly in terms of time, of managing a team. In technology the manager needs to be technically involved, to monitor what the team is doing, to work on short-, medium-, and long-term strategic and budgetary plans, to deal with incidents, and even to be able to evaluate employees' performance. That's why, in addition to direct people management tasks, it's important to have time to work on technical projects, even if small ones, to keep up to date with all the technological changes.

The situation can get worse for the overburdened manager if their manager doesn't have much more experience than them, something that is common in companies where the workforce is very young. In such cases, it causes professionals to work 12 hours shifts to try to take care of everything. This situation is not sustainable, and they will eventually look for better opportunities and leave the company. Burnout and team inefficiency can directly result from an imbalance in team composition. In any company, you want to increase productivity and reduce management overheads.

There are several ways to reduce management overhead, but one key point is undoubtedly rationalizing the number of direct reports under a leader. There is a limit to how many people a manager can directly lead, but it needs to be a variable number that works well across the board. Many people have been studying this issue for a long time, and my goal here is to approach the problem only from my practical experience.

The span of control is too small

When I began at Stone, the technology area expanded rapidly due to the company's success in payment systems. At that time, they had 2,000 tech employees, of whom approximately 500 were managers. On average, the teams comprised four people, with a span of control of three (one manager for three direct reports). There were even more extreme cases where a manager only directed one other employee.

Very small teams are a good solution, as they reduce the management burden on managers, who have fewer people to manage and, therefore, fewer administrative tasks. But the correlation is not direct. The very fact that someone is a manager already implies a series of duties that demand dedicated time—in other words, there is a management overhead inherent to the position.

Another complicating factor at Stone at that time was the leaders' lack of seniority. Inexperienced managers spend more time and energy leading the team. With a huge number of managers and most of them being inexperienced, a quarter of the technology team's 2,000 employees' time was consumed by management tasks, and managers were practically unable to make any technical contributions. All in all, this meant a huge productivity loss.

Anyone who manages technology teams is bombarded with problems every day, regardless of the team size. There are all kinds of conflicts, from operational ones, such as someone needing to use an unresponsive system or a new feature that is buggy and is hurting users, or the need to advocate for more budget to finish the project, to personal problems of employees, such as a disagreement between two employees, or inappropriate behavior that requires correction, or an employee facing a serious problem in the family.

When the person in charge of the team has the experience and know-how to act, time and energy are saved. An inexperienced manager, on the other hand, will probably need to pass on part of their burden to their own manager or mentor, consulting them on what to do. In doing so, they multiply the management overhead. If the inexperienced manager does nothing, the team conflict or technical problem goes unresolved and hampers productivity.

To balance the technology team at Stone, I adopted a strategy that combined increasing the seniority level of the managers with giving them a wider span of control. This way, the company would have larger teams led by more experienced leaders.

The first step was determining that no junior person could be a manager. In every case where someone led a team in this stage of their career, an effort was made to merge teams, which reduced the number of leaders and increased the span of control.

The next step was to see which management employees would like to become individual contributors. This alternative didn't exist in such a structured way before, and it presented itself as a new career prospect that we needed to publicize and encourage after formalizing the Y-career path. This allowed us to slim down the bloated management arm a little more. None of the reassignments negatively affected employees' compensation.

The strategy for increasing the span of control adopted concrete targets **based on data**. As the employees and management were all mapped, it became possible to set gradual targets for increasing the ratio of direct reports to leaders. We started with a span of control of 4.5[18] direct reports per manager and set year-on-year targets, which over two years, raised the ratio to 6.5 direct reports per manager, which is much more appropriate.

Of course, there are always exceptions to this ratio: a super-specialized job may require a high level of control and a very small team, and an activity with a very standardized process would allow the manager to command more people. In any case, I believe that the ideal number of employees for someone to manage in the technology area is between 7 and 8, sometimes reaching a maximum of 10.

The reorganization at Stone aimed to increase the team's overall performance, make employees feel more productive, and allow managers to have more time to mentor and motivate the employees.

One way to measure results for this type of plan is to periodically check the company's turnover rate. At Stone, we saw positive impacts, with a drop of more than 20% in the turnover rate, reaching a better-than-market average in 2023 reflecting greater employee satisfaction and more stable teams.

---

18. I am using realistic but illustrative figures, which should not be taken as Stone's exact reality.

Seniority pyramid

The balance between experienced and inexperienced professionals is crucial for a technology team to function in the way I envision: an inspiring and stimulating work environment where employees feel productive and have room to be creative. This will increase the team's overall efficiency and allow innovation to happen.

Experienced and talented professionals on the technical track can make a huge difference to the company, as they are able to take a broader view of setting up platforms that will give strength and agility in the long term. It is possible to surround such a person with collaborators who will complement their abilities and help them to have an even greater impact on the company's activities.

Senior professionals on the management side, on the other hand, can lead large teams with a clear direction, conveying stability to their direct reports and helping consolidate the organizational culture. Both on the technical and on the management side, key leaders are also multipliers of the platform mindset for the rest of the team.

I was very lucky to have started my career at IBM, a company where the management of new researchers was very structured, with extremely experienced leaders who gave me solid direction. If I had started in a garage start-up with a group of people the same age as me, I would have had a very different professional experience.

Leaders with corporate experience and a strategic vision built up over many years are better placed to allow room for creativity while still providing technical guidance. Problems of team misalignment in technology are incredibly complicated because they can lead to duplicate code, systems more complex than strictly necessary, and potentially significant technical debt. People in technical and managerial leadership roles need to be alert to these deviations.

I know that a company's resources will always be limited, and I also know that the technology team tends to "eat up" a large part of the budget of companies in all industries. What I advocate for, and one of the reasons why I'm writing this book, is that investing in top talent is worthwhile. It's an investment that pays off most of the time because by hiring a highly efficient senior manager, this person will reduce management overhead across the board, attract other equally efficient people to the company, and so on.

I argue that we should consider the organization as a pyramid with a narrow top. You must preserve the pyramid format, with levels of seniority distributed throughout the different layers, so as not to fall into the trap of having a huge team in which no one really knows where they're going.

When you identify a team with too many junior professionals, you can create a seniorization plan like the one we did at Stone. There, the model determined a point on the career path to serve as a threshold between someone considered junior and senior and controlled the proportion of new hires. If the company previously brought in 15 junior employees for every senior hired,[19] this could be changed to 3 juniors for every senior until the desired pyramid format is achieved. I understand that this is a long and gradual process. However, it also must be intentional, and already planned out when defining the annual budget and the opening of each headcount.

In the first decade of the 2000s, Google decided to establish a laboratory in Brazil by acquiring Akwan, a company originated in the university environment of UFMG (Federal University of Minas Gerais) and headed by Professor Berthier Ribeiro-Neto. As the head of the Brazilian lab, Berthier maintained a cautious hiring approach, focused on quality and a balance between experienced and inexperienced professionals, even as the team grew rapidly. With this cautious approach, Berthier led the growth of the Brazilian lab and positioned it to work on crucial innovation projects for the Google ecosystem. Today, the lab has grown tremendously, and I attribute its success in part to their manager's ability to preserve a balanced pyramid of seniority.

Management capacity

In addition to the parameters of direct reports per manager and the absolute number of managers, another

---

19. The figures are merely illustrative and do not reflect Stone's reality.

factor that needs to be considered is the number of layers in the organizational structure. Organizations with too many layers can suffer from communication problems because directives must go through a long "word of mouth"—or keyboard-to-keyboard—process, from the top commanding layer down to the lower ones.

Reducing hierarchical layers improves communication and makes the company more agile. Still, the automatic consequence of doing so without reducing the number of employees is to increase the span of control, creating larger teams, which brings other losses. Determining the number of layers and the average number of employees per manager establishes the company's management capacity.

It would be great if this were a simple calculation, but it is not. As I pointed out when I spoke about the exceptions to the ideal size of the span of control, there are variations in relation to seniority and the number of hierarchical layers, depending on the activity and the strategic position of the organization. This is not something that can be imposed evenly on the company. But it's good to control these parameters with global metrics and metrics by area, so that measures are taken based on data, and not just subjective perceptions.

For example, a research area, which works on medium- and long-term innovations, may require very senior professionals. Areas such as customer support and internal systems, despite being important, tend to have a higher proportion of junior employees, larger teams, and more junior leaders.

A metaphor to represent the management focus is the construction of a house. While on the one hand management needs to monitor the laying of the foundations and plumbing very closely, the tiling service does not demand as much attention because, should something go wrong, it would be much easier to replace the tiles than to replace the plumbing or correct a faulty foundation.

Once you define that the average span of control of a company is 10, and that you don't want more than five layers of hierarchy, the maximum number of employees will be 11,111—1,111 of whom will be managers. Without monitoring, it's quite possible that at some point the leadership will realize that the company has 20,000 employees and only 500 managers. Or that, on the contrary, the number of managers is inflated by 5,000 for a total contingent of 15,000 employees. There are various types of imbalances that must be avoided or corrected when detected.

The signs that the organization is out of balance will appear clearly in the metrics derived from HR systems, but they can also be seen in other ways: employees go into burnout, turnover increases, the number of code changes delivered to production drops, incidents happen more frequently and take longer to mitigate. The DORA (DevOps Research and Assessment) set of metrics can help detect some of these imbalances.

Being data-driven and thinking intentionally about team structure is one of the key activities of the technology leader. The organization is always evolving, and we need to keep a growth mindset, open to learn new parts of the business and optimize the team to increase productivity and reduce management overhead. Actively optimizing team structure is part of the platform mindset approach described in this book.

Resource allocation

In strategic decisions about organizing a technology area, an increasingly significant variable is the possibility of hiring external services, rather than keeping teams working "in-house." Outsourcing certain activities is inescapable, especially after the popularization of cloud computing. Outsourcing is a way of controlling the number of employees, bearing in mind that you will still need a team to interface with the service provider, although potentially much smaller.

Outsourcing should be considered when deciding the team size and structure. It makes sense to reserve resources for developing in-house the systems that compose the company's competitive edge and the areas with potential for disruption, as well as to contract services readily available on the market for areas seen as commodities. It doesn't make sense to build a system to detect whether a server is failing because cloud computing services, such as Microsoft Azure, Amazon Web Services, and Google GCP, already do this well and efficiently.

On the other hand, systems that need to scale, such as Stone's payment systems, could represent a very high resource commitment if contracted out to a third party. It may be more advantageous to keep their development and operations in-house because this would allow them to plan for productivity gains and control future costs.

Outsourced platforms go on the company's balance sheet as an operating cost, also called OpEx, short for "operation expenditure." In general, personnel costs are considered operational. In a technology company, however, we can view part of the personnel costs as CapEx, which are capital expenses, or how we classify investments in the company's tangible and intangible assets.

People's salaries can be considered CapEx when working on creating a new product (a new system is a company asset, just like a new building). Capital expenditure provides for a depreciation rate over time, like when a company buys a batch of computers. The investment is expected to depreciate over the years as the machines break down.

OpEx and CapEx spending are reported on different lines in companies' balance sheets and are analyzed by investors to assess the company's health. For the technology leader, all the CEO (Chief Executive Officer) and especially the CFO (Chief Financial Officer) may want to know is whether the spending in the area falls under OpEx or CapEx.

Prabhakar Raghavan, who hired me at Yahoo Research, once told me: "For the rest of the company, everything you do is a line on a spreadsheet." It must be a part of a technology leader's intention to be familiar with these concepts—which I won't go into detail here—to know how to justify the investment in equipment, projects, and personnel to management and investors. Planning includes having answers on the expected return on investment (ROI) right on the tip of your tongue. It's a structured vision that facilitates communication with the rest of the company. In this regard, the technology leader needs to work in partnership with a finance professional who acts as a controller or a chief of staff.

The budget for the technology area is generally considerable, and the CTO or leader of the area cannot promote the eternal growth of their team because that would be impossible. The goals of tech leadership should be aligned with those of the CFO.

What technology is aiming for is something the finance area also likes: gaining operational efficiency and reducing costs. This is done by increasing automation, producing more scalable software, and reducing technical debt. What allows these advances to happen are largely the gains in human material, what management does to direct the results of employees' daily work. If management efforts work out well, there is a reduction in personnel costs, because fewer hands are needed to run the day-to-day operations. This reduction in operating expenses leaves more resources to be reinvested in the form of CapEx.

A sustainable team for the company's current number of managers and a balanced seniority pyramid make up the best conditions to assess its appetite for continuous investment in technology. The decision of where to inject more resources is linked to its innovation ambitions, as detailed in Chapter 4. When putting money into long-term innovative projects, the aim is that this money will, over time, be transformed into concrete products, services, and efficiency gains that make up the company's day-to-day business, the operating cost of what is known as business as usual (BAU). Whenever there are efficiency gains, there are resources left over that can be reinvested in projects with a more distant horizon, perpetuating the innovation cycle.

Thus, the technology team will always have a big part of its contingent working on the maintenance and operation of current systems, a part on medium-term projects aimed at optimizing processes and platforms, and a part on long-term structural and exploratory projects.

Organization of teams by project or area

Between 2011 and 2013, I was on a team that worked on the infrastructure of the Google.com search engine. The organization of that team was very fluid. Let's say we had a group of 50 people, with five managers. Each manager had a team of 9 people. Mixed into this pool of 50 people were technical leaders and the five managers also had much technical involvement. I, Marcus, who reported to team A, didn't necessarily work on the project led by the team A manager. I could be working simultaneously on the project led by someone from team B and on another project led by the manager of team C. My project colleagues could come from teams A, B, C, D, or E. It was a matrix organizational structure because we knew that projects were more temporary than people.

Even though I was always working for other technical leaders, when it came to going through my performance review or discussing my vacation, I had to speak to the A team manager, my official team. In such an organization, the managers in the pool need to be reasonably well informed about what is going on in all the projects so that they can, for example, evaluate the performance of their employees.

As Colin Bryar and Bill Carr describe in their book *Working Backwards: Insights, Stories, and Secrets from Inside Amazon*,[20] Amazon established very early on the practice of creating teams composed of people from various areas who dedicated 100% of their time to a specific project, such as the development of the Kindle. These are called single-threaded teams, and the single-threaded leader at Amazon oversees a team dedicated to developing that specific product. Each of these teams is independent enough to execute their projects without dependencies from other areas of the company.

Amazon's very particular model was born from the realization that the need for communication

20. *Working Backwards: Insights, Stories, and Secrets from Inside Amazon*, by Colin Bryar and Bill Carr (St. Martin's Press, 2021).

and approval between areas was holding back the company's projects. Instead of striving to improve communication, Amazon decided it was better to eliminate the need for communication altogether. This assemblage of mixed teams is similar to the product development structures adopted more recently by agile methodologies.

The advantage of these models is that they allow choosing the leaders with the most suitable experience and expertise for each project. These teams are also more likely to form spontaneously, in response to the members' motivations and professional interests, to address a new business opportunity they identify.

In general, however, it is much more natural for any company to adopt a traditional organization determined by area of activity or function. In a company like Stone, the banking area is divided into sub-teams, such as credit cards and core banking. However, this does not preclude hybrid management models, which adopt more flexible organizations within smaller teams.

From a macro perspective, it makes sense to divide teams by area of activity at the level of the entire company. But within a smaller organization, with 50 people or less, for example, there is room for interchangeability between projects. Individual contributors usually fit into formal teams for evaluation and administrative purposes but can be allocated independently to specific projects. The free transit that ICs have between teams and projects is one of the advantages of establishing the Y-career path.

Management rites

Even if there is an emphasis on dividing the team up into short-term projects, it is advisable to manage each employee's work and career progression in the long run. I'm in favor of one-to-one meetings (1:1) with my direct reports every week. As I pointed out in the previous chapter, meeting once a week allows for cancellations without too much damage to the regularity of the meeting. Regular weekly meetings create a good rhythm, which makes them objective and doable in half an hour's conversation.

As Camille Fournier points out in *The Manager's Path*, it's not enough to maintain an "open door" policy and tell employees that leaders are available to assist them whenever necessary. Only by actively seeking information can one monitor each professional's career and work, especially considering the distance imposed by remote work and the introverted personality of many people working in tech.

Meetings make up a considerable part of the management overhead, so it's important that they are conducted rationally and efficiently, without losing the personal element that makes them a unique tool for developing the company's human resources.

Streamlining meetings means using them for the few processes that will work better synchronously rather than asynchronously—through written communication, for example. Preparing agendas so that people arrive with already thought-out points to discuss are good practices that should be encouraged.

Agreeing on what will be discussed asynchronously before the meeting can make it much more productive.

Another habit I've adopted and recommend is skip-level meetings, where an employee meets with the person who manages their manager. This type of meeting can be held at the employee's request or on a routine basis, individually or as a group. At Stone, I meet with all my skip levels every 60 days.

It's a valuable type of meeting to increase transparency within the company and to detect problems and opportunities. Perhaps the manager identifies a talent that deserves to be closely monitored or learns that their direct report—the manager that is being "skipped" and who is therefore not present at the meeting—is unable to keep up with the team's work, due to being overloaded or maybe due to any other less noble reason. Direct contact between managers and their skips increases supervisory power and gives employees more confidence that there is room for them to express their ideas and concerns.

On the other hand, skip-level meetings should not be requested by people trying to gain direct access to the top levels of leadership just to stand out, as this strategy is very unlikely to work. It should be borne in mind that senior managers are, by nature, very difficult people to impress.

It is not healthy for an organization when there is a widespread impression that subjective criteria and personal favoritism are the norm and that some employees are taking advantage of loopholes in the systems for their benefit. This kind of perception creates a terrible working environment. People management processes need to be very clear, and management needs to be closely aligned with them.

I often say that a company's management team should work like an algorithm. If someone asks any of the hierarchical leaders: "What does an employee need to do to be promoted from level B to level A?" the answers should always be similar, demonstrating that they all have the same promotion standards, and see the company's processes through the same prism. Achieving this uniformity among managers is a laborious task, and the constancy of rituals is an important element for leaders to form a cohesive and aligned group.

Tools instead of processes

To ensure uniformity in management tasks, ideally, we should create automated systems and written protocols. A well-described career plan and a performance evaluation system that works as a collaborative tool based on that plan are examples of practices that leave the company less open to favoritism and injustice. The rules are explicit and integrated into the tools, and neither management nor employees need to spend energy trying to interpret them.

Written protocols are still subject to each manager's interpretation. When protocols are transformed into tools, the room for questioning is greatly reduced, and the frequency of thoughts such as: "How lucky I am to work for person X, because they are 'nicer' and promote their direct reports without demanding as much as person Y" is reduced as well. If we want to build high performance teams, we cannot afford this kind of inconsistency.

Tools and processes act as safeguards to minimize misinterpretation. At the highest levels, promotions should be conferred and approved by a group of managers, rather than by a single person, to reduce the risk of bias. The more consistent the behaviors are, the more confident people will be that they are part of an impartial organization with clear standards, which will give them more mental space to carry out their duties efficiently and creatively.

Creating people management tools, monitoring the area's data and basing strategic decisions about the team on the analysis of this data, ensuring clear communication, balancing the seniority pyramid, and controlling management overhead are team organization measures that support a fertile environment that fosters innovation.

PRINCIPLES FOR TEAM ORGANIZATION

- Define and manage the span of control (number of direct reports per leader) and number of hierarchical layers.
- Reduce management overhead, the overload of team administration tasks so that managers can also do technical work.
- Balance the seniority pyramid so that there is clear guidance to junior employees and the organization does not lose quality standards.
- Balance in-house development and outsourcing.
- Allocate personnel resources taking different innovation horizons (short, medium, and long) into account.
- Reduce uncertainty and subjectivity by turning processes into tools.

Chapter 4
Innovation management

When I was at Microsoft, I often heard variations of the same question from people on my team: "If we have so much to do in Azure, why have a team focused on quantum computing?" The answer to this and other questions involving experimental or seemingly futuristic projects is simple. Every company needs to invest in longer-term learning horizons if it doesn't want to risk losing its competitive edge.

I remember Satya always saying: "I need my technology leaders to see around the corner." I share this same philosophy because nurturing environments that encourage the emergence and development of ideas is the only antidote to the innovator's dilemma, as defined by Clayton Christensen. According to the Harvard professor and author of the bestseller *The Innovator's Dilemma*,[21] companies that focus their attention and resources only on what they already do well are susceptible to becoming obsolete in the long run, with the emergence of new technologies or being overtaken by the more agile competition.

---

21. *The Innovator's Dilemma: When New Technologies Cause Great Firms to Fail*, by Clayton Christensen (Harvard Business Review Press, 2015).

With these issues in mind, in this chapter, I will discuss how to apply the platform mindset to keep the flame of innovation burning bright in terms of organizational structure, processes, and even engineering tools. I'll also give some examples of how certain innovations have emerged in the companies I have worked for and try to establish general concepts that can be applied by engineering leaders in their own teams.

Scientific method as a tool for innovation

Those of us accustomed to today's personalized and dynamic advertising on the internet may not know or remember that, at the beginning of the 2000s, ads were practically static and not necessarily targeted to users' interests and tastes. It wasn't uncommon, therefore, for someone to be reading about the results of sports events on a news portal and see ads for winter coats.

This was the context when I joined Yahoo Research in 2005. Prabhakar Raghavan, who led the initiative to infuse the company's processes and products with the scientific method, which he called "making the company science-proof," tasked us with looking at the issue of serving better ads in an exploratory way.

To lead this research, Prabhakar recruited Andrei Broder, whom I mentioned earlier. Yahoo Research was situated in Silicon Valley, and once a week, we flew to Burbank to work with the team responsible for the ad systems. Most of the team members came from Overture, which was acquired by Yahoo in 2003.

Our goal was to understand the existing system and find ways to evolve it, both in terms of the relevance of the ads being served and the cost and availability of the system. I spent about a year and a half making this trip on Wednesdays with Andrei and Vanja Josifovski, another young engineer at the time who would later become the CTO of Pinterest and Airbnb. On the other days of the week, we explored and tested possibilities.

At a certain point during those visits to Burbank, we realized that the existing system wasn't producing relevant ads and wasn't stable, and that the way ads were served needed a radical change. It didn't need a simple incremental addition to what was already in use but a break from the norm. Since we had already been working on some ideas, this was an opportunity to put something new into production to validate our research hypotheses.

I'd like to take this opportunity to say that this turning point, promoting the system we developed in our research to Yahoo's ad production system, was only made possible because of the support we received from Qi Lu and David Ku, the executives responsible for the company's production systems and visionaries in the way they supported innovation.

From conversations, observations, data collection, and analyses, we concluded that we needed ads that captured people's interest and were more contextual and complementary to the content displayed—a concept that is quite widespread in today's ad-serving systems but was novel at the beginning of the millennium.

The new area of **computational advertising** came from this melting pot of well-educated and curious people who had been hired by Yahoo with the freedom to work independently and creatively and were willing to solve complex problems.

Andrei Broder is a master at inspiring the team and creating an environment that fosters collaboration and innovation. I worked with Andrei for almost ten years on various projects, all of them successful. He never gave us a fixed deadline for any deliverables. In all these projects, the team was always highly motivated, and a lot of this enthusiasm came from Andrei knowing how to stimulate us. When we started working on computational advertising, many of the problems he proposed were new, such as "How can we represent that people who like Formula 1 might be more susceptible to Rolex ads?"

We solved this problem of Rolex-Formula 1 ads using information retrieval techniques, representing these concepts as documents and computing similarities between them. Therefore, the documents representing the query "Formula 1" and the Rolex ad could be understood with semantic information, such as "luxury" and "eccentric," and these standard terms would weigh up in the similarities shared by both.

We had no idea back then that this new approach to ad serving that we helped develop at Yahoo would completely change how people advertise on the internet. I give Andrei enormous credit for that.

Andrei is a very talented computer scientist and theoretician who can also share the world's most unusual life stories. He is hilarious and fun to be around, using every opportunity to present a vision of possible paths that no one else can yet see on the horizon and charting how to generate practical results from this vision. With all this, he was able to awaken our own creativity for engineering projects, from the lab at IBM, where I first met him, to Yahoo and Google, where so many of us went in the following years.

Anyone who works directly with me will probably have noticed that I am a disciple of the "Andrei school" of project management. Talented and capable people are naturally motivated, so putting pressure on them to meet fixed deadlines usually doesn't positively affect the project. On the contrary, unnecessary demands can generate stress, which is one of the great enemies of a work environment conducive to creativity.

The Google model and the Microsoft model

A lot of corporate literature is dedicated to analyzing whether top-down or bottom-up management is better for fostering innovation and producing successful products. That's not the goal of this book. I'd like to share here some of my experiences creating technology teams and a technology culture that promotes innovation.

One of the technology companies with the most significant impact in the first decades of the 21st century was created with little focus on management. Google was founded in 1998 with the initial intention of making a search engine, and to do so, it decided to hire the best people on the market.

These hires were made so fast that, at one point, the technology managers had such large teams under their command that they couldn't even tell who their direct reports were (the problem of high span of control, as discussed in Chapter 3). When I worked there, from 2011 to 2013, the organizational structure was still very flat, with few hierarchical levels.

With so many good people working there, it was no surprise that many ideas were always springing up. This was especially true considering that there was an official company policy allowing people to devote 20% of their time to any project they wanted to get involved in. The seeds of interesting projects, such as Google News, came from these more untethered moments. Andrei Broder refers to this management model as "let the flowers bloom."

However, there came a time when an average of only three or four people worked on these small projects that didn't necessarily contribute to the business. Google then began implementing a slightly more top-down management hierarchy, consolidating some of these small projects. However, they kept alive the Google principle of giving autonomy for new ideas to circulate between teams, with the help of engineering tools standard to the entire company.

At Microsoft, however, there has always been a more significant focus on organizational hierarchy and planning, with well-thought-out scenarios that must be implemented. From the outset, managers had, and still have, the mission of figuring out what problem needs to be solved and how to organize the teams to deliver solutions to that problem in the next planning cycle. It's a model focused on the efficiency of directing what you already know must be done: the engineering leader plans the project in detail and allocates people according to the needs of what will be developed in that cycle.

Despite this apparently more rigid management structure, when I was at Microsoft between 2013 and 2022, I had room to create, and the rigid planning process did not lock up the team. It certainly happened this way because we had teams with a few employees above what was strictly essential to carry out day-to-day tasks, understanding that engineering needs to operate with a certain amount of fat to balance out short-, medium-, and long-horizon projects—or H1, H2, and H3, in the nomenclature popularized by the McKinsey consultancy at the end of the 20th century. The team's fat allowed us to explore new ideas without compromising our current projects. It was, therefore, a hybrid between top-down and bottom-up management.

Hybrid path to innovation

Promising ideas that were prototyped and developed in addition to the scenarios already defined in planning eventually need to become official projects. For this to happen, they need to be approved by management. If someone develops a new cool service, one way or another it must be submitted to the technical judgment of one or more leaders before it goes into production.

As I mentioned, Google engineers had a lot of autonomy to validate an idea and put it into production without needing higher-ups' formal approval. This was possible because the engineering processes and tools guaranteed that the changes would not negatively impact customers. For example, if someone had an idea of how to make the search stack cheaper, with a lower service cost for Google, without changing the quality of the search engine's response, the change would be approved through the usual unbureaucratic code review process. Now, if the idea reduced the cost relatively, but impacted the quality of the results shown to the users, we had to discuss the tradeoffs with the leaders responsible for the system.

The reality is that this need for approval from management didn't happen very often at Google because we knew that the answer would be "no" if the changes impacted users. We ourselves were very concerned about the quality of our work.

The way we operated in the event of a change that directly impacted our customers, such as an API change, usually involved creating an MVP (minimum viable product) to put into production for only a small percentage of users, so we could validate the idea. There was also a lot of experimentation and A/B testing, where a new feature was tested with some users while the other portion continued to interact with the old feature. This way, we could find out whether people liked it or not.

When I mention these situations from my years at Google, they shouldn't necessarily be seen as the ideal formula for all companies but rather as a snapshot of the organizational structure that allowed the company to create such innovative services. One of the pillars of this approach was the fact that Google was able to attract a team of experienced and professionally mature engineers right from the start. There was confidence at the top of the company that no one would waste their time experimenting with trivial or useless projects.

There were also few management processes because there was a lot of control over how the engineering tools were used. Each team was responsible for a part of the codebase and the only one way to change it was getting approval from the original creator. The approval process was integrated into the platform, preventing anyone from making changes and putting them into production without them being tested and approved.

Another difference is that the projects I worked on at Google then were for consumers. When it comes to a business-to-business (B2B) product, the process for putting any new feature into production needs to be much more rigorous, as it necessarily involves changes to APIs that require development on the part of customers to be adopted.

My experience with B2B products was at Microsoft Azure, where we carefully planned all the features and established project timeframes that could be communicated to customers. However, even in B2B projects, it's always possible to make internal changes that don't impact external interfaces, such as improving the performance or the availability of a system.

The comparison between these two management models, bottom-up with team autonomy, and top-down with more rigid planning, leads me to believe that what ends up working in many technology companies is a **hybrid model** between the two. Not everything in an organization must be planned so rigidly, with well-defined delivery deadlines, but not everything can be left completely up to the developers.

Research x production

Experimentation doesn't have to be limited to huge companies able to afford a fancy research lab. When Google attracted and hired researchers who were working on search engines, the company didn't allocate them to its research division. They worked directly in production teams.

When production and research work separately, it is often the case that researchers, the only ones with a "license" to exercise creativity and innovation, would develop a prototype and hand it over to production. Oftentimes this is a recipe for failure, because the innovation is not shared with those responsible for developing, maintaining, and operating the systems.

Acting as the link between research and development has become one of my main specialties, and this started at Yahoo Research. Despite being a research lab designed to be independent, we soon became involved in production systems, trying to validate our ideas gradually, rather than delivering a finished prototype to the engineering teams.

The following diagram shows how research and production teams can mix. Neither research nor production should be 100% of the team. During the experimentation phase, the research people dominate the team, but from this point onwards, it is essential that there is some involvement from the production engineers. In the implementation phase, the ratio is reversed, but research people are still involved.

Figure 3: Technology team composition throughout the innovation cycle

EXPERIMENTATION   PRE-PRODUCTION   PRODUCTION

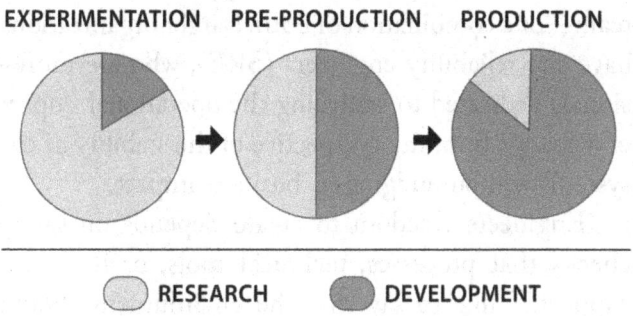

RESEARCH   DEVELOPMENT

Creativity with safety

At Stone, we often discuss what kind of corporate culture we should embrace for experimentation, especially since there is no department dedicated exclusively to research. Can people test whatever they want?

The answer is no. The government highly regulates the area in which the company operates. Stone even has a product committee with professionals from various areas, including legal and regulatory affairs, to ensure that every product that goes into production meets all the legal and regulatory requirements.

Another concern is the stability of systems, including cybersecurity, performance, and availability. For example, before a feature goes live, the cybersecurity team validates that there are no loopholes that would leave the company vulnerable.

We also must consider that however small a change may be, it can lead to instability. To minimize this instability risk, safeguards need to be put in place. This **technical validation structure** can be either an automatic testing process or the approval of a specific team—or a combination of both. Larger organizations have site reliability engineers (SRE), who are professionals dedicated to analyzing the operational impact of a change from the perspective of the stability of the systems without judging its business impact.

Engineers' freedom to create depends on safety checks that processes, technical tools, or both can support. I like to say that the environment should enable **fearless execution** from the part of the engineers—if they do anything silly, which is bound to happen, there will be safeguards in place to protect the production systems.

Even when there are regulatory restrictions, such as "every credit card now has to contain a chip," it's still possible to exercise creativity. The technology leader can always organize their team to allow innovation to flow and influence product definitions.

At Stone, we have adopted ambidexterity between the product and technology teams. This means that for each of the company's products, such as banking, there is an engineering leader and a product leader who are peers and whose teams work together. We believe that there is **product creativity** and **engineering creativity**, and they can be synergistic (see Figure 4).

Figure 4: Product and engineering innovation

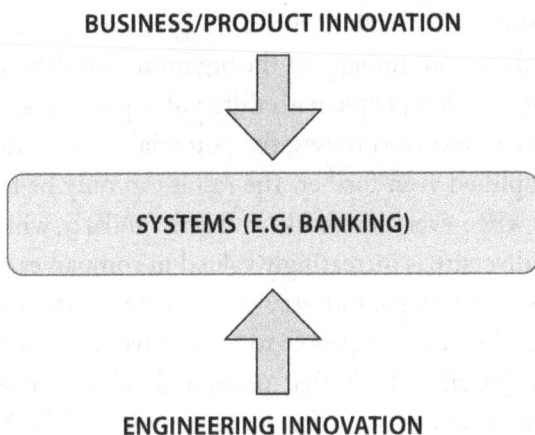

BUSINESS/PRODUCT INNOVATION

SYSTEMS (E.G. BANKING)

ENGINEERING INNOVATION

Many people may need help understanding what engineering creativity is, but it is expressed in how a system is operated, how its deployed into production, or how the cost of running a system is calculated and balanced. Many engineering innovations are entirely independent of product innovations, and the way it works best is when they come together to boost the quality of a service. Doing real-time fraud analysis is an engineering innovation that can enable product innovations—for example, instant credit.

Ideas may come from all directions

When you think a company can have hundreds or thousands of employees, it's impossible to believe that only a few leaders at the top will produce the seeds of innovation. I may be Stone's CTO, but 2,000 people work there in technology. It wouldn't be wise of me to waste the minds of 1,999 people because I'm the appointed leader.

There are certainly several bright, innovative individuals spread throughout the organization. When you bring together people with different experiences, backgrounds, and worldviews, the potential for innovation is amplified even further. The result can only be better than when everyone thinks and acts similarly, which is why diversity is increasingly valued in companies.

What Google did very well was to create mechanisms that allowed people to be creative with the help of engineering tools that promoted collaboration. It was very easy to prototype an idea, build an MVP, and test it in production. At the same time, an engineer who wanted to start a new project had to exercise their powers of persuasion with other colleagues to "sell" their project and gather support to move it forward. The reality is that it's not enough to have a good idea, it needs to be supported by others. For this to happen, the company must have mechanisms in place.

Companies with a more top-down approach tend to have fewer of these mechanisms available. It is essential to enable people lower down in the hierarchy to access decision-makers. With this access, employees feel like they will be able to contribute something new and be more motivated to bring ideas forward.

People in leadership roles must open channels of communication so that all employees feel that they are listened to and that they can contribute ideas and innovate. Ideas can come from anywhere in a company, so I see as my responsibility to keep the flow of ideas open.

At Stone, in my Ask Me Anything chat with everyone who joins the company, I always make it clear that I'm open to hearing innovative ideas. Institutionally speaking, it's common for companies to have demo days, where people can present ideas to the whole company, and hackathons, to think problems through and validate an idea in a fun and stimulating way.

Some companies even allow their employees to raise money internally for their ideas, creating mini venture capital initiatives within the company itself. All of this is valid, and each technology organization must search for solutions that are compatible with its culture.

One way to make room for ideas to emerge in practice is to invest in creating that excess fat that gives teams resources to think outside the box. This fat allows for the eventual creation of new medium- (H2) and long-term (H3) projects.

My experience as a technology manager shows that not every employee wants to dedicate their time to new ideas: there are those who prefer to work on projects that have a direct impact on the business. Most people fall into that category. Ther are few people, however, whose creative side is more pronounced and who only feel fulfilled when they have the freedom to create new things.

Good leadership knows how to identify how each member of their team works best and takes advantage of their strengths. A balanced team needs a mix of people working towards different planning horizons.

Projects with different impact horizons

Innovation is essential for the existence and endurance of technology companies, so having professionals working on more experimental projects does not constitute a concession, but rather a business necessity. One way to encourage ideas to flourish is through formal programs in which, after a certain period in a department or project, the employee is encouraged to do something different in the company.

New ways of looking at old problems often generate surprising solutions. All the companies I've worked for have programs in place where, if an employee wants to change jobs within the company, no one can stop them, as long as they have a good performance review and the manager for the area where they want to work welcomes them.

Having common engineering tools makes it easier for people to move between projects and departments. Imagine what a waste of time it would be if employees must learn a new programming language and new tools every time they join a new project.

New projects don't always attract many people, because they have less impact on the present and are often seen merely as experiments or speculations. Sometimes it takes years before the area has a product to deploy into production, but it's part of the investment in innovation to support teams like this.

When I was at Microsoft, I started a project in a completely new area, power management in data centers, with only one engineer working on it, Nithish Mahalingam. Nithish spent years working alone with zero visible impact from what he was doing. After a few years, however, that area took off, gained prestige and importance, and generated a significant impact for the company. We built an area from the ground up until we reached a very high number of millions of dollars of impact.[22] At that point, of course, everyone and their grandma wanted to join the project.

---

22. No, unfortunately my compensation was not proportional to the impact the project generated.

It's essential for leaders to encourage their teams to work on exploratory projects that have neither present impact nor glamour yet. I understand that each person's appetite for professional risk is different (and depends on their life circumstances at the time), but we must encourage career moves and point out the advantages of sometimes creating something entirely from scratch, like Nithish did.

Believing in your ideas

One of my first projects at the IBM Almaden Research Lab, where I worked from 2000 to 2005, was a research prototype I built with Vanja Josifovski, a colleague as junior as me at the time, whom I mentioned earlier in this chapter. Before working together in computational advertising at Yahoo, we also collaborated at IBM. One day, we ran into a technical problem.

We had to store and process very large XML documents, containing, for instance, all the income tax returns for some US states, in IBM's relational database (DB2). The difficulty was that these documents would end up being larger than the available memory and, consequently, impossible to process by being materialized—the prevalent technique at the time.

Vanja came up with the idea of processing the document in a streaming fashion. To minimize memory usage, the file would be processed bit by bit. This way of processing XML documents using a streaming algorithm was innovative at the time, and no XML processor operated like this.

Nobody put much faith in or gave much initial support to Vanja's streaming solution, but I found it interesting and agreed to work on this idea with him. Even though we were discouraged by our direct managers, we went ahead. Vanja and I knew that we worked in a research environment that offered a certain amount of career autonomy. We also knew that we had to manage, at the same time, our obligations to other projects that were under our responsibility. So, we quickly put together a prototype without any direct endorsement, but being careful to validate our ideas to make sure we weren't wasting time on something ineffective.

Once other people saw that it worked, our prototype became a supported project, and we started working on it officially until it became the ultimate solution to the problem of processing large XML documents in the database. We discovered an engineering innovation (a novel streaming algorithm) to enable a product innovation (processing large XML documents in a relational database). It was a success story involving two very junior employees, with direct managers that opposed the idea, although they didn't stop us from trying.

I like this story from the beginning of my career, because it illustrates how it is possible, even without a leadership position, to do work that has an impact on the organization. Vanja and I believed that we could do it, but we also agreed that we wouldn't waste hours of work on a speculation that wasn't based on good technical judgment. We were also aware that, just like any experiment, it might not have worked, but the company wouldn't punish us for trying. Even younger professionals can act maturely when faced with a challenge.

On the other hand, in a corporate environment with inexperienced and professionally or personally immature employees, it can be costly to open the innovation gates. If everyone were allowed to work on whatever they want, without basing their proposals on technical knowledge, the situation could soon become chaotic. Career paths are a good point of reference: after a certain level of seniority, engineers could be granted more autonomy to decide how to prioritize their projects. Hackathons and demo days are a great vehicle for employees to showcase their projects and prototypes.

Conscious investment in a balanced mix of H1, H2, and H3 horizon projects prevents the business from settling for their bread-and-butter, which will inevitably be copied by the competition. As I mentioned in the introduction to this chapter, this type of investment is the antidote to Clayton Christensen's innovator's dilemma. Companies whose teams work almost entirely on H1 projects may want to foster innovation, but this becomes practically impossible. How can anyone devote themselves to new projects when they are completely busy with day-to-day work?

Firefighters cannot create

There are teams that only work on "putting out fires" or managing day-to-day problems, instead of having time reflect on how to improve their systems. If everyone in your organization is acting like a firefighter, there's no point in saying that there's room for people's creativity to flow.

Technical debt, incidents, and malfunctioning systems also end up compromising a team's or individual's ability to be creative. In general, the company's systems need to run without frequent incidents and complications, so that people can free up more of their time to think beyond the existing problems. Consistently dealing only with problems ends up generating frustration and burnout in employees.

I know that it's common for startups employees to work days and nights, because the team is small and everyone is rushing to produce first versions of products. But that doesn't mean that this way of operating should be perpetuated as the company's working model once it matures.

I often tell my teams at Stone that, unless there is a high-priority incident, for instance, impacting our credit card processing machines, everything we do today can be deferred to tomorrow. Emergencies must be seen as one-offs, not day-to-day work, and it's up to the leadership to spread this culture.

Not every apparent good idea works

Since I mentioned the successful XML processing project Vanja Josifovski and I developed at IBM, I think it's fair to bring up another personal story in which an idea didn't work out.

In 2011, when I was working with the late engineer Luiz André Barroso[23] at Google, I came up with the idea of using a new RDMA (Remote Direct Memory Access) communication library to speed up query processing on our search engine, using a graph processing strategy instead of the traditional RPC (Remote Procedure Call). I convinced Luiz to allocate three engineers to support me on the project for about six months. In the end, we didn't manage to achieve what we wanted, despite the very positive intermediate results. A long time later, I learned that the ideas generated in this project were recycled and are now in production in other Google systems.

In this kind of situation, you need to remember that in engineering (as well as in other areas) you always learn from your mistakes. We should adopt a growth mindset and investigate what went wrong, knowing that future adjustment of ideas is always possible. For example, maybe what doesn't work for project A will work if applied to project B, and there will be no time wasted trying out ideas that have already been proven to not work well.

It goes without saying that we all want to get things right more than we get them wrong, and the level of risk appetite varies greatly from one professional to

---

23. Luiz had a unique vision of innovation, formalizing the concept of "roofshots" and the reaction between roofshots and "moonshots," described in the Roofshot Manifesto (available at https://fontoura.org/papers/barroso.pdf).

another. I have a rigorous academic background and was trained to apply the scientific method both during my academic training and later in my industrial research career at IBM and Yahoo. Having understood that it is part of the scientific process to propose a hypothesis and test it to reach a conclusion, I have worked over the years in the corporate world quite comfortable with the concept that many of our tests will fail.

The scientific method is gradually being integrated into the culture of technology companies, and people are becoming more and more comfortable with the process of experimentation and gathering data to test new ideas and products. Agile work philosophies, such as the "lean startup" methodology first described by Eric Ries,[24] establish that there should be a simple and quick way to initially test an idea and gather data before moving on to a large-scale implementation.

Amazon, whose management methodology is described in the book *Working Backwards*, which I've already talked about, has also had a big influence in business environments with the concept of managing with data and metrics, including how and why to promote ideas and products that boost the business in the light of customer needs.

---

24. *The Lean Startup: How Today's Entrepreneurs Use Continuous Innovation to Create Radically Successful Businesses*, by Eric Ries (Crown Currency, 2011).

Short and long races

Business as usual (BAU), or a company's H1 investments, is what currently generates value for the company, whether it's the Office suite or Microsoft's cloud service. There needs to be budget available to operate these products at the highest level. Many people don't understand that once a system or service is in production, it still must be operated, because improvements and maintenance will always be needed. Every time someone uses Google Meet, for example, they are using a service that requires constant monitoring so that it runs smoothly on Google's servers.

The operational cost of keeping these products running for any technology company is high. However, a company that only focuses on BAU and doesn't invest in innovation, has no way of growing beyond a certain point and will end up being left behind. The capital market sees the value of a technology company precisely in what it has besides its bread-and-butter. What the market is betting on is that the company will be able to innovate on what it already offers today. It's as if the market is confident that, overnight, Google may create another YouTube or Meta may launch another platform as popular as Instagram.

For a technology company, this potential translates strongly into building engineering platforms that allow it to scale the size and complexity of future offerings, as I will explain in depth in Part II.

Engineering platforms allow companies like Uber, which originated as a passenger transportation service, to expand is scope to include the delivery of goods, food, and freight, or Netflix, which started out as a mail-order movie rental company, to move on to video streaming producing its own content.

Companies like this created robust platforms to fuel growth, which in turn enabled new product offerings and nurtured an environment of innovation. These were intentional decisions, guided by a platform mindset, that diversified the business and prepared it for what lay ahead.

For great innovations come to fruition, an unquestionable ingredient is the creativity that lives within the people who make up the company, the artists. There may be funds to create and a positive environment, but the protagonists need to be there to make it all happen.

MEASURES TO ENCOURAGE INNOVATION

- Encourage teams from all areas to test ideas and solutions in a more experimental way.
- Cultivate a culture that is open to listening to and promoting ideas from all levels (this can be institutionalized in some way, such as demos days).
- Promote the philosophy that creativity can and should be fostered in any sector of the company, whether in research or production.
- Establish different working horizons in the short-, medium-, and long- term (H1/H2/H3).
- Adopt common engineering tools and integrated systems to facilitate exchanges between departments and greater diversification of ideas.

PART II
TECHNOLOGY

Chapter 5
Reusable platforms—the "bikes"

When my wife, Renata, died in 2006, Andrei Broder and Prabhakar Raghavan, my Yahoo managers, offered me a sabbatical to deal with the emotional burden of that moment. At first, I was reluctant to accept it because I thought working would at least give me some continuity in the face of the loss and radical changes happening in my life. I felt that I would be even more groundless without work, given how much it fulfilled me.

However, as at that time I had already been in the United States for more than eight years, away from my parents, a few months after Renata's death I decided to accept the proposal to take a break for nine months. As I was interested in studying yoga and meditation, I spent the first four months traveling around India and immersing myself in this learning—I even studied a little Sanskrit at the time, although I've already forgotten everything, unfortunately. I lived most of the time in the southern Indian city of Mysore, where I met yoga students from all over the world. At that time, I hardly did anything related to technology, but I had contact with a very different reality and many friendships were born there.

The following five months were spent in Rio de Janeiro, living with my parents and teaching a graduate-level course at the Pontifical Catholic University of Rio de Janeiro (PUC-Rio) about how search engines work, alongside Professor Eduardo Laber,[25] a subject we named "information retrieval for the web." Those months at PUC-Rio were vibrant—I loved teaching, and it was rewarding to spend time with several professors who were so crucial to my education.

To this day, I maintain a strong relationship with PUC-Rio, which culminated in the creation of StoneLab, Stone's research laboratory situated there, which opened in 2024. I will talk more about StoneLab in Chapter 9.

The months in India and Brazil did me a lot of good and confirmed my initial feeling: despite missing Renata, developing technology and working on the platforms we were building at Yahoo were a very significant part of my identity. The time away gave me more drive to return and contribute to building something innovative at work.

When I returned to Yahoo, I was quickly able to get back to being productive, since we had built foun-

_____

25. Eduardo Sany Laber is an Associate Professor at PUC-Rio, and his research interests include machine learning and theoretical computer science. More information available at https://www.inf.puc-rio.br/blog/professor/eduardo-sany-laber/.

dational platforms for computational advertising that served to accelerate our innovations. I was already familiar with these platforms, so it wasn't difficult to continue thinking about other applications based on them. It was also a special experience to arrive after almost a year away and discover that the team had grown substantially, and the new people who had joined were already publishing papers using the platforms I had helped build. It was nice to see all that progress and realized that the platforms were a big part of it.

This chapter focuses on platforms, and I started with a very personal story precisely to illustrate how integral this subject is to developers' daily lives and my philosophy of life and work (and therefore the title of this book). Building platforms has enabled me to understand the value of investing time, energy, and resources in a foundation built to support scale and innovation. You must wonder how some of our creations leave a legacy far beyond us.

Over the following few pages, I'll show you how platforms function as the fundamental building blocks from which products that will make a difference to a company's current and future business will emerge. Platforms help make engineering more agile and help position technology companies. I also explain how platforms impact team culture.

Hierarchical platforms

A platform is a system that can be used in more than one context. For example, Google initially wanted to build a search engine. But it didn't just develop a search engine, it instead, built a reusable platform. Google engineers wrote a system for code review, a generic system for testing, a system for deploying code into production, a generic communication library, a generic data storage framework, and systems for monitoring applications and servers, among several others. By designing all this in a generic way, Google was able to reuse what they already had at hand as a starting point for its new offerings, such as YouTube or Google News. Thus, when a new idea came along, it wasn't necessary to write everything from scratch, because the basics were already in place.

A computer system is composed of a hierarchy of components connected to each other, such as Lego pieces.[26] Lego building blocks can be viewed as the most basic of platforms, and you can build anything using them. Once you have used them to assemble a basic structure, like the chassis of a vehicle, such structure can be reused in different contexts. You can build ambulances, Formula 1 cars, trucks, and tractors faster by reusing the generic chassis. In other words, it is much easier to reuse a platform than to have to build everything you want from scratch.

There are platforms at various levels. Platforms that validate code or deploy binaries into production are engineering platforms that support a wide range

---

26. I use the word Lego as a generic term derived from LEGO® building blocks.

of systems. Platforms for managing customer data or billing clients are more specialized product platforms—they solve the specific problems of a given area. Product platforms are built using the engineering platforms as foundational blocks.

Let's say a company builds an application for delivering ready-made meals for dogs (the meals are delivered to the dog owners; the dogs just eat them). The technology team works tirelessly to build the entire application, but shortly after they are done, the company decides to expand the business to the lucrative area of dog grooming. Now that it has two different businesses, where will the customer information be stored? Who is responsible for this information?

To solve this issue, the company creates a common customer information management platform. With this platform in place, the company realizes that it also needs a common scheduling platform. So, little by little, their team builds reusable platforms, which will not only be used for all the current offerings but will also work for future offerings, such as the even more lucrative business of custom-made dog outfits. The technology team can also rely on third-party cloud services as part of the platforms to support their activities.

Cloud computing services function as generic outsourced platforms. These services, like Microsoft Azure, Google GCP, and Amazon AWS, run on servers scattered in data centers around the world. This means that instead of needing to buy and manage a server to run your systems, cloud providers do that for you and for millions of other customers. This is feasible only because cloud providers have platforms that work at scale, from the server's fault management systems, to the logistics systems that can repair or replace specific parts of a server, to the systems that ensure that the servers run in a sustainable way, with a low carbon footprint.

With these generic platforms at their disposal, a small startup is free to focus on its core business. If the startup doesn't want to use a cloud provider, it would need to consider a series of technical issues, such as where it will run its systems, who will restart them if failures occur, and how to keep the systems connected. This is obviously possible, but it implies excessive costs not linked to the core product of the business itself.

The strategic decision about what systems to build in-house falls under the umbrella of engineering leadership. The key, however, is to think in terms of reusable platforms. At Stone, for example, we decided to invest in a core banking platform generic enough to serve all our customer segments (from micro merchants to large businesses).

The platforms behind the business

When we talk to investors about Stone's business, we always say that the heart of the company is different from the banking or the payment systems, even though these systems are highly robust and complex.

Our strength resides in our ability to **create any scalable system**, be it for banking or payment or anything else. It's the engineering platforms we've built that make the difference to the business, because they're what will allow the next system and the next business to be built, in addition to providing scale and improving the systems already in production. It is this infrastructure that will make it possible for Stone to decide to operate in another country or another market segment, without having to start from scratch.

An image created by João Bernartt, who used to be Stone's CIO, and which I like very much, has become mantra of Stone's technology transformation: instead of spending all our energy running hard to win every "marathon," we are also investing in building bicycles, which will allow us to compete more efficiently in many future races. The "bicycles" are our platforms. In the early days of the company, we won several races with our own grit, running as fast as we could. This led Stone to build a series of large-scale systems, but siloed, without a common framework or uniform engineering guidelines.

Since 2022, however, we have shifted our focus to building bicycles, investing in the long term and in a common computing framework, which will give us more business agility. In a marathon, it's worth "wasting" the time needed to build long-lasting bikes—the races are long, and those bikes will help us win not only the current races, but also the ones we cannot yet foresee.

I'm convinced that technology companies need generic and reusable platforms to have a competitive edge. With these in place, the different teams have plenty of time to try out new things, let their creativity flourish, make mistakes, learn from them, and produce transformative innovations, both in engineering and in business.

A generic super platform

Sometimes, innovation arises from the curiosity of a person or a group looking for answers to questions that seem unrelated to their daily work. Innovation can also come as a direct result of the constant search for more efficiency, something that is very present in the corporate world and in the daily lives of technology companies.

When I was working at Microsoft, some colleagues and I wanted to understand how to improve the efficiency of Azure, our cloud computing service. Cloud services provide the abstraction of virtual machines, but on the provider side, the equipment is very much physical. Virtual machines run on servers in data centers. If the physical computer fails, the cloud automatically provides the user with another virtual machine that, from the client's viewpoint, works just as any computer would. We started by asking ourselves how clients used our services.

Our goal was to gain a deeper understanding of how these virtual machines' resources were being used: were they being used to run one-minute calculations, such as a spellchecker, or to run long-term services, such as the Bing search engine, which had to run 24/7?

Cloud services provide virtualized computing services, whether for storing photos from a personal cell phone or computing delivery routes for a logistics company. We also need network connectivity to enable these tasks. The cloud abstracts all of this (storage, computing, and connectivity) and returns a combination of these resources to clients according to their demand.

We wanted to develop a system to predict how resources are consumed in these several scenarios. For example, if a client asked for a new virtual machine, we had no way of knowing if it was to run a photo backup system, which would use little computing and more storage, or an artificial intelligence calculation with complex matrix multiplications, consuming a lot of computing power. These are very different uses of the cloud, which are reflected in very different computing resources needs.

We concluded that we should create an artificial intelligence system to predict how cloud resources would be used. We built a service called Resource Central,[27] a system that consumed utilization records,

27. The scientific paper on Resource Central was published in 2017 and is available for consultation at https://www. microsoft.com/en-us/research/wp-content/uploads/2017/10/Resource-Central-SOSP17.pdf.

or logs, from the several cloud services to derive predictions about resource consumption.

When we decided to work on this, Ricardo Bianchini, a researcher at Microsoft (and who is now a technical fellow and corporate vice president of Azure), and I recruited Eli Cortez, a young engineer at the time (who is now senior director of engineering at Azure), to join us. We followed the evolution described in Figure 3, starting out as more of a research team and gradually becoming more of a production team.

When we developed Resource Central, we still needed to find out which applications would need predictions about the use of Azure computing resources. Many teams might want to understand cloud usage behavior, but we had no idea how many, so we built Resource Central as a generic platform. We could have chosen to build something to understand cloud resources only in terms of energy demand, or to calculate a specific storage utilization metric. Instead, we prioritized building a completely generic prediction system because we were aware that we wouldn't be able to predict all future usage scenarios for the platform.

I bring up the case of Resource Central because it's a project that we started from scratch, with us drawing on a whiteboard, and which turned into a system that today serves hundreds of applications at Microsoft, far beyond what we could have imagined when we conceived it.

The initial team was the three of us, Ricardo, Eli, and I, plus Anand Bond, from research, who worked with Eli. When I left Microsoft in 2022, Eli already had a team of more than 20 people just developing Resource Central, and Ricardo had a research team working on new versions of algorithms and prediction models. The work evolved both in applications and in scientific publications, including a paper on why we made the Resource Central's architecture so generic.[28] In the end, the main scientific innovation of the project was not even the ability to make predictions, but the way in which we managed to organize the platform in a generic way, decoupled from most systems, giving it immense potential for adaptation and reuse.

I'd like to point out that nobody told us we had to make Resource Central, and nobody even knew that not being aware of cloud resource utilization was a problem at the time—our small group anticipated a problem and created the platform. This is a practical example of the "let the flowers bloom" mindset in an environment that fosters innovation.

Here, I'll take the opportunity to include a short story about the characters in this project and our interconnected paths. Ricardo studied at the Federal

---

28. "Toward ML-Centric Cloud Platforms," Ricardo Bianchini, Marcus Fontoura, Eli Cortez, Anand Bonde, Alexandre Muzio, Ana-Maria Constantin, Thomas Moscibroda, Gabriel Magalhaes, Girish Bablani, Mark Russinovich, *Communications of the ACM*, February 2020, Vol. 63 No. 2, pages 50–59.

University of Rio de Janeiro (UFRJ), Brazil, then did his doctorate in Rochester and went on to become a professor at Rutgers University in New Jersey. When I was doing my post-doctorate at Princeton, between 1999 and 2000, also in New Jersey, Ricardo already had an excellent reputation as a professor.

I had the opportunity to visit Rutgers and introduce myself to him. I was just a student, so I was very happy to meet him and grateful for his generosity in welcoming me. We chatted for a short while in his office. Imagine my surprise when, almost 10 years later, I bumped into him at Microsoft. I worked on search engines, and he worked on systems research, so at first our activities had nothing to do with each other. But, at a certain point, I began to have more interactions with Ricardo, and we formed this partnership where he was the research arm of my team for several projects that produced a lot of impact on the company.

I first met Eli during a conference in Indianapolis in 2010—the same conference where I first met Luiz André Barroso, who was giving the keynote speech. Eli was a student at the Federal University of Amazonas (UFAM), Brazil, and later joined Microsoft Research as a post-doctorate student. When he told me he was considering changing areas, I invited him for an interview. I ended up hiring Eli for the small experimental team I was setting up in Azure, and from then on, I've followed his incredible professional development. He's someone who started at Microsoft at a very young age and today has a team with enormous influence in the company. I'm very proud to have actively participated in Eli's professional development.

Platforms and architectural design

A particularly inspiring author for me is the architect and mathematician Christopher Alexander, who discusses the architectural design process in works such as *Notes on the Synthesis of Form*.[29] According to him, the construction of buildings and cities follows systematic patterns. Something like, there always must be a "bathroom" and a "bedroom" module in a home, and, at a lower level, the "bedroom" module should have windows. At higher levels, such as in a "community" module, there should be leisure areas for the inhabitants, such as parks.

These reusable modules encapsulate rules for building functional spaces, from cities to buildings and houses. Similarly, when developing systems, there are platforms that are already built and tested and that encapsulate complex business rules. These platforms guarantee the correctness and cohesion of ideas.

If you look at the diagrams of Stone's platforms, you'll see a lot of generic engineering modules until you get to the payments system, which is something more specific. We even joke that, if everything else goes wrong, we could turn Stone into an electronic

---

29. *Notes on the Synthesis of Form*, by Christopher Alexander (Harvard University Press, 1965).

games company, because so much basic infrastructure is already in place, such as client data management, mobile application, and machine learning platforms, that we wouldn't have such a long way to go if we wanted to change industries completely. Platforms can also be extended and adapted to different scenarios.

We can think of a "swimming pool" module, with certain design structure and variations in formats and finishes. One fine day an architect thought of how to extend that basic module to create an innovative, borderless pool, which came to be known as the "infinity pool." The "infinity pool" still preserves the properties encapsulated in the "swimming pool" module—it must have a foundation, hold water, and so on. The same is true for software platforms, which are not static and must evolve over time.

Platform stacks composed of layers

Can Lego bricks be seen as limiting because the standard piece is square or rectangular? Perhaps. But even with this limitation on shapes, you can build anything from a spaceship to an ice cream parlor to a princess's castle out of Lego pieces. The limitation is not restrictive in the sense that it doesn't prevent anyone from creating whatever they want, however they want.

When you build reusing larger structures made with Legos, there is obviously a bigger restriction. However, if the larger structure meets your needs, the restriction has a positive side: a builder who starts from pre-built boat structures to build an entire marina can save time when compared to building boats from scratch. The pre-built structure encapsulates the complexities of boat design, allowing the builder to focus on the marina.

Platforms are essential to achieving product agility—it is much faster to build a marina once you have the boats. Platforms allow the company to innovate without wasting time reinventing the wheel. There's no need to recreate an entire computing framework every time a product or service idea comes up. The applications and services built on top of generic platforms implement your business differentiation, such as Stone's core banking system (see Figure 5). These modules can be designed to work in a generic way, such as a payments module that works for Stone's various businesses, from integrations designed for micro-entrepreneurs, who rent surfboards on the

Figure 5: Layers of a computing platform

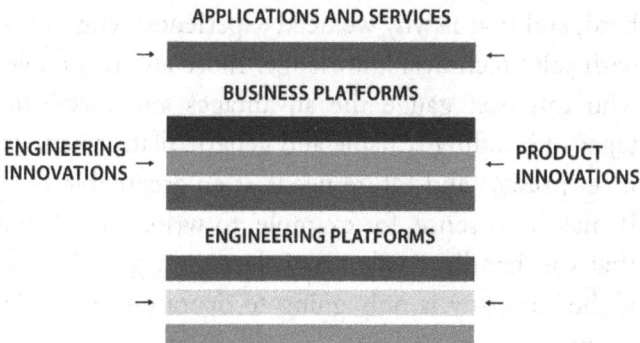

beach, to larger companies that need more sophisticated solutions to pay their suppliers.

Platforms should also be built with growth in mind. Let's say you build a platform that will support 10x your current number of customers. When you go beyond that scale, it may still work, but not at its maximum efficiency. The response time for some services may no longer be satisfactory. With the number of customers increasing even more, that platform no longer works. It becomes an ungainly wagon, and we must either improve it or rewrite it. If you don't do anything, you'll have to deal with technical debt that will continue to accumulate over time.

Unlike Lego pieces, which are easy to fit together, computer interfaces are more complex, so they require modules that communicate well with each other. If the way the interfaces fit between the different levels of a platform is more perennial, changing it is easier. If the design is not conceived in a scalable way, you must constantly rethink the abstractions.

Architecting these layers and evolving them is hard, and that is why we need experienced engineers with solid technical knowledge. Those are the people who can best gauge the advantages and disadvantages of building scalable and generic platforms that meet present and future needs at an acceptable cost. It makes no sense, for example, to write a platform that can handle a volume of data on a global scale if the company is only going to operate in a single country.

In the book *Working Backwards*, the authors tell how Amazon's product catalog was built as a monolith. With only one system, the product catalog, anything that needed changing, required adaptations to that fixed structure. As Amazon grew, many teams had to change that system all the time, and any modification risked making it unstable, affecting customer usability. That's why, at a certain point, the company concluded that it needed to break the monolith into smaller components. This was a project that lasted several years.

If companies had a crystal ball to get all the abstractions of a platform right from the start, it would be very simple. Since that doesn't exist, all you can really do is rely on scalable and reusable platforms with simple abstractions.

Let's say a company has a platform for predicting fraud that analyzes all its customers, taking 15 hours daily to complete the process. The problem is that the company has grown 10x since the platform was built, and the platform now requires 150 hours to complete the same task, which obviously can't be done in a day. As it starts to incur losses, the company is forced to rethink the way it predicts fraud.

The same logic of changing the amount of data analyzed can apply to search engines like Google and Bing, which must constantly reassess the relative quality of all the pages on the web. There are processes that analyze this overnight, looking at the relative relevance of each page and site, such as Facebook, *The New York Times*, the neighborhood newspaper, or a college student's blog. These pages need to be ordered and ranked. But back then the number of pages on the web was one and now it's something completely different, and the platform still must continue performing this relevance analysis anyway.

Platforms provided by third parties

The engineering and business platforms available on the market are becoming increasingly better, but they don't solve 100% of a company's problems and still need to be adapted and integrated into the company's systems. Despite the enormous versatility and usefulness of cloud services, it would be delusional to think that the cloud has everything a company needs to support all the technical aspects of their business.

The company must have an engineering team to take care of the integration, checking if the interfaces with the cloud services are working properly and if the company's use of the cloud is as efficient as possible. The vehicle chassis is ready to be used, but someone must connect it to the rest of the car and check if it runs efficiently and doesn't come loose at some later point.

Therefore, as I've outlined in previous chapters, the technology leader has to understand what is a commodity and what is a business differentiator for the company, and where and how they are going to allocate their human and financial resources—this is an important part of the platform mindset.

A company should focus on features that differentiate their business. At Stone, our differentiators include the banking and payment applications, the credit card processing machines, and the systems that run on these machines. If the cloud didn't exist, we would need to hire more developers to implement the computing platforms that we currently outsource to the cloud providers. By using third-party platforms, we can focus the attention of our engineering teams on competitive differentiators for the company.

Most technology companies won't compete with Amazon, Microsoft, or Google, and they have no reason to do so. They must look for a different space from what these giants already occupy, and then innovate in that direction. There are, of course, engineering ideas that can be differentiating business ideas for companies, such as using artificial intelligence to predict server failures. A successful company implementing this idea could probably end up developing services that would be adopted by cloud providers. And in that case, the company's competitive edge would be an engineering innovation.

Platforms, and especially internal platforms, tend to be invisible to customers. It's common to have one group of engineers building a service using version 1 of a platform, while another group is already thinking about its version 2, maybe to gain more scale and efficiency. Improving a platform from version 1 to 2 can be very complicated and involve many technical aspects. It is like changing the wheels of the car while the car is still running. The change should happen without impacting the services that use the platform.

Normally, developing and improving platforms is more long-term than delivering a new version of some functionality to clients. When planning the work for the teams, it is important to consider different cycles. Platform development may require longer cycles and less rigid delivery dates.

We'll see later in Chapter 7, when talking about incidents, how the organization should deal with the implementation of these new versions of platforms to minimize any unforeseen events. The general idea is to avoid major changes all at once and to implement incremental changes over time. This limits the potential impact to only micro-incidents, rather than macro-incidents. An example of this is to deliver a new feature for just a slice of the population, rather than all users.

Any problem with the deeper layers of the platform stack may have a large impact on the other layers, while a problem in a higher layer is less contagious for the whole. Using a house metaphor, a problem in the foundation may impact the entire house, while a problem in the hardwood floors may just require a few boards to be changed. Generic platforms are great for solving complex issues for several teams and have undeniably a multiplier effect, but if they fail, they also have a negative multiplier effect. A platform for developers, used by the entire company to write and test code, is a typical case. If it stops working, it becomes a huge problem, impacting the work of all engineers in the company.

Platforms as a catalyst for innovation

At Stone, the applications that customers interact with run in the cloud, and from the company's point of view, the cloud platform is a commodity. On the other hand, if we were a cloud provider, we would need, for instance, experienced people to predict how long virtual machines will run for. By doing this, we could make better use of the system's resources and make the cloud more efficient, as we did with Resource Central at Microsoft.

Microsoft thinks about their cloud systems, but doesn't need to innovate in banking applications, while Stone must move in that direction. Intel innovation is on their processors, because they want servers to consume less energy, last longer, and be more resilient, but for their customers, the processor is just a basic building block that allows them to achieve their business goals.

Platforms stacks have many levels of abstractions, and the investment on innovation will be on the level of abstraction that is the company's competitive differentiator. For Stone, these differentiators are the financial platforms created to make life easier for our customers.

We also have a risk assessment platform, which is extremely important for the operation of the business. It predicts the risk involved in providing a loan to a customer, for example. The engineering innovation there could be finding a way to reduce the time it takes for the platform to identify whether a person is a fraudster or not. If the system, instead of taking hours to process this computation, took just a few seconds to give an answer with a good level of certainty, Stone would be able to grant credit instantly to its customers at very low risk, which would be a huge competitive advantage.[30]

Even though it is a consumer of third-party services, a company like Stone still must build its own platform to deal with the challenges of detecting and preventing money laundering and cybersecurity, which are critical issues specific to its business. These are examples of platforms that require both engineering and business innovation.

If, to resolve the hypothetical example of instant credit, we could use a new service that provides more data, such as a person's history of paying their utility bills, it could unlock a business innovation that would produce a more assertive risk assessment platform. If, in an even more radical example, we had authorized access to absolutely all financial data from everyone on the planet, throughout their life, we could probably build a very precise risk assessment platform.

In that case, we would have to think about how to process all the data efficiently, since the volume of information would be much larger. It would be a trade-off between more cost and processing time and the assertiveness of the result. With a challenge like

---

30. This is just a hypothetical example, not related to any of Stone's product strategies.

this, the hope is that an engineering "artist" would discover an algorithm to speed up the process from days to seconds by compressing the data or using more efficient indexing techniques. A platform innovation like this could enable a huge range of new products. Figure 5 shows that the platform stack can benefit from engineering and product innovations.

When to invest in platforms

Whenever you use a third-party platform, it never fits like a glove, and you must make adaptations. There are pros and cons that need to be considered, such as cost or processing speed. The external platform can do 90% of the work, and the other 10% may need to be implemented internally. For example, the platform could be from a foreign company that can determine quite accurately whether someone is a fraudster but may need to be integrated into the local credit bureaus.

Platforms can be developed either because a company has a unique need or because it wants to innovate with some new product, such as offering instant credit to its customers. There is no point in developing an expensive platform if there is no clear business application. If a platform that takes three days to generate a customer's credit score is computationally cheap and does what the company needs, it is probably good enough. However, if another credit score platform is 10% more expensive, but produces better results, a switch may produce a good return on investment (ROI) for the company.

The technology leader and the CFO should constantly analyze where and how best to invest the company's financial and human resources. Before any development starts, we must analyze its cost, how long it will take, whether there is an externally available platform that solves the same problem, and whether it will be possible to develop it to be reusable and generic to solve other problems.

It's always good to remember that reusable platforms may avoid several independent applications to be developed, sooner or later, by different teams, containing technical debt. Avoiding this kind of duplication is one of the main objectives of the initiatives I've implemented as technology leader at Stone. My focus has been unifying different technology areas and platforms and thinking of standardization as the backdrop to improve collaboration and efficiency.

Platforms evolve

According to architect Christopher Alexander, for everything in the world there is an evolutionary design process that leads to a "perfect" design. In his book *The Timeless Way of Building*,[31] he discusses design standards, timeless architecture, and the limitations of concept and style. He believes that there is an evolution of forms that works almost like a mathematical process: if you haven't yet figured out how to fit your

---

31. *The Timeless Way of Building*, by Christopher Alexander (Oxford University Press, 1979).

Lego bricks together, you need to study more to get there, since a solution is always possible, and there is only one perfect solution—the timeless one.

I mention Christopher Alexander again because he is a very influential author in the world of computer science, which took from him the idea of architectural patterns as the basis for software architectural patterns.[32] My doctoral advisor considered Alexander's *Notes on the Synthesis of Form* to be one of the most inspiring books he had ever read, and my Ph.D. thesis was on reusable software architectures, also closely linked to Alexander's concepts.

That said, I don't advocate striving for Alexander's perfection and definitive form. There can be more than one way to build a successful system. Over time, as technology evolves, new ways of building may also become available. Any product created today comes with a mobile-first mentality. However, before the advent of smartphones, that was unthinkable. What wasn't a reality in 2007 suddenly became one.

I understand the theoretical limits of computer science, but we have seen great changes in how we interact with computer systems, even using natural language as an interface. The artificial intelligence revolution is significant, and we thought it would take longer to get

----

32. The book *Design Patterns: Elements of Reusable Object-Oriented Software* (Addison-Wesley Professional Computing Series), one of the most influential books in software engineering, was inspired by the work on architectural patterns proposed by Christopher Alexander.

to this stage where the interface with the computer is not necessarily programmatic anymore. This certainly opens other design avenues not available in the past.

When natural language really does become the preferred way of interacting with machines, will anyone still want to have millions of apps on their cell phone like they do today? It's hard to foresee what will happen, but the trend is for computing to become even more widespread and for all devices to be connected through specialized interfaces, focused on the task at hand. Instead of using a generic computer to write a book, you might use a better, more specific device, perhaps activated by voice. My guess is that we'll become more and more like the Jetsons.[33]

Luiz André Barroso used to say that imposing restrictions on platforms is an invitation for all the brilliant minds in the world to think within the confines of these limitations: how to use that interface better, how to make that component more efficient or more scalable. For Luiz, a good example was the instruction set architecture (ISA) of the x86 chip developed by Intel in 1978, which runs on every computer on the planet. Luiz always said that when Intel defined those instructions, people immediately said that they were too restrictive; however, they became the standard and generated countless innovations.

---

33. *The Jetsons* is a series of animated cartoons produced by the Hanna Barbera studio in the United States between 1962 and 1963, and then again between 1984 and 1987.

Imagine if every researcher studying computer architecture had their own standard—it would be much more difficult to evolve the architecture because instead of working on the same problem, they would be working on many different problems. Luiz believed that standard restrictions often enable innovation.

In the following chapter, I'll show how the standardization of engineering tools is another key ingredient to enabling innovation in technology companies.

PRINCIPLES FOR IMPLEMENTING REUSABLE PLATFORMS

- Design generic platforms for reuse, such as Lego blocks.
- Develop a layered platforms stack, with engineering-focused and business-focused layers.
- Evaluate building platforms in-house and outsourcing them.
- Enable teams to innovate, by focusing on the company's differentiated offerings.
- Understand that generic platforms have a multiplier effect, positive and negative.
- Consider how technology innovation, such as smartphones or AI, may impact the platform stack.

Chapter 6
Engineering tools

The year 2011 was a turning point in my life in many ways. On the personal side, I was remarried and recently became the father of two little girls. At work, I joined Google, and for the first time, I found myself facing the challenge of working purely in production engineering and not in research.

My relatively short three-year stint at Google was very significant for my professional career. There I found a way of working that was very much in line with what I believed in. The engineering tools were standardized, and much of the communication between people and teams was done through these tools and the code itself, reducing uncertainties and inefficiencies.

Many senior managers don't understand the importance engineering tools have in facilitating collaboration between developers and between teams, and consider the day-to-day processes—the programming languages used, the testing environments, the validation system—just minor issues that need neither investment nor planning, due to the huge range of tools available on the market and to the variety of internal processes that each team creates to execute their own projects.

In this chapter, I will show how these concepts are mistaken and how tool standardization has the potential to bring huge productivity gains to the company. I'll explain how sharing repositories, a culture of testing and code review, and the integration of guidelines into tools enable engineers to produce the platforms that will differentiate the company and allow them to be efficient and innovative.

Shared repositories and information at your fingertips

One of the aspects that shaped Google's work culture almost from the start was its investment in engineering tools. These tools were built to make life easier for developers, who had at their fingertips everything they needed to write, test, and deploy new code into production. There was no need for so many processes between people, because the tools themselves simplified actions and no one wasted time thinking about how to perform tasks in the engineering lifecycle.

The engineering tools were so standardized that there was only one way to put the code into production, and this approach dictated the collaboration strategy. One of the main elements of this system was the existence of a common repository that brought together all the company's code. What's more: it was indexed. Every time someone wrote code, it became searchable by keywords. After all, it would be very strange if Google didn't have a good search system for its code. Programming languages were also standardized and, while I worked there, basically only two programming languages were used by the whole company.

Once I was working on a project and all that was needed to finish my feature was to create a monitoring panel showing how the servers were connected. I used a graph to represent the information spreading in real time, so that if there was an incident, someone would know at which point the execution plan was. I then did a search on Google's internal code search tool to find out if there was already any code to do what I wanted. And I found it.

As there was commonality in the way the code was written, with common code standards and libraries, it was very easy to read what other people had written. When I tried to use that module, I saw that there was a small problem with the code and, as it was a simple thing, I found a solution. All I had to do then was look in the repository directory to find out who was the author responsible for the module I was using. I wrote the necessary modification and submitted the code review request, without even knowing from which team or country the "owner" was from. Within a few hours, the person responsible reviewed my request and approved it, and I was able to check-in my change to production and use it. There was no need for any personal interaction, because all communication was done through the code review tool. I went on to work on other projects without ever having met the owner of the code or, better yet, without having needed four meetings to resolve the issue.

If the code wasn't indexed and searchable, I would have had to ask people if there was a graph modeling library. Some companies have hundreds of repositories, and it's impossible to search them one by one. Perhaps I would have looked externally to see if there was anything open source, which many people end up doing because at least there is a search tool for that. Suppose I still haven't found what I wanted. In that case I'd have to write the code myself, and I could be duplicating or triplicating something that already existed in the company and spending time and energy in the future on testing and maintenance. The process might have taken a week or more rather than a day.

Of course, no tool is infallible. Even with the Google tools, there could have been a delay if, for example, the person responsible hadn't been so readily available. However, with this type of integration, the possibility of the process working in a truly agile way is greater. Also, **collaboration through tools** allows work to be done asynchronously and remotely, as I'll go into more detail later.

Efficient production line

Engineering tools are what software engineers use daily. To explain briefly for those who aren't familiar with this: a developer spends his life writing, compiling, and testing code. Sometimes they also sleep and do other things, but that is rare, so let's ignore that for now. When they are satisfied with the tests, they submit both the code they have written and the tests to be reviewed, ideally by technically experienced engineers. The reviewer can and should make comments and ask for changes, and when everything is right and approved, the code goes onto a conveyor belt we call CI/CD (continuous integration, continuous delivery), which is a mechanism for automatic validations of that code, checking, for example, for security loopholes, or that the formatting is done correctly.

After going through this validation environment, the code goes into production, i.e., it enters the repository—before that, it was only on the developer's machine—and is integrated into it. This is a process that directly affects developer productivity, and it can be slow depending on the amount of automation and bureaucracy involved. The goal is for most internal processes to be automated, and one of the points that has the most impact on this is the repository—whether it's one for the whole company, one per team, or one per group or sub-group. What happens when these repositories are scattered around the company, in large numbers and in a disorganized way, is that there is no easy way to visualize them, and the code they store.

There are many reasons why there is an overabundance of repositories. This is how the situation usually unfolds: the company is making, say, a credit card processing system, and the developer creates a repository for the system's code. As it grows, a new director is hired to run the banking system. This person thinks that the system will never use anything from the existing credit card processing system, and creates another repository, building a completely independent system.

A company like Stone can have numerous repositories like this, which is often explained by the need for speed. The smaller the repository, the faster the tools work. If the process is getting bogged down, the developer feels an impulse to create a new, separate repository, without integrating it with the rest of the codebase. You may even have several repositories, but if you want developers to collaborate and reuse code, there needs to be a search tool that works as if there were only one.

A very senior engineer friend of mine once said to me: "Marcus, when you told me how easy it was to work collaboratively at Google, I thought I understood, but it wasn't until I went there and tried their tools that I realized what a huge difference they make." People don't understand how much productivity improvement these tools give developers and how they enable collaboration.

I remember when Google decided to build Google+,[34] a project, which didn't succeed, to try to compete with social networks like Facebook. The new social network needed a way of doing search queries, and I, who worked in Google search, followed the project, limiting myself to observing the requests for approval of changes by the Google+ team, which was reusing Google's existing search infrastructure.

---

34. The social network called Google+, or Google Plus, was created in 2011 and deactivated in 2019.

If these code collaboration tools hadn't existed, it would have been necessary to call in a senior engineer, with the impressive title of Google+ senior search architect, to talk to the search team and design how the Google+ search feature would work. They would hold a meeting, discuss things in an abstract way, draw boxes and arrows on a whiteboard, then put together a slide show, generating a huge management overhead that was avoided. Google+ may not have worked as a project, but at least the internal search proved to be solid—it was done very quickly and worked in a scalable and cheap way, without us needing that senior architect getting paid to draw complex systems on a whiteboard.

Companies that don't have these tools in place are forced to create mechanisms involving people, meetings, and processes, with a resource commitment that would be much lower if the issue of engineering tools had been resolved. I'm not against meetings; I'm even a big fan of 1:1 meetings, as I explained in the chapters about people management. They should, however, be reserved for matters in which interpersonal exchange is valuable and irreplaceable.

Tools for learning and culture change

In Chapter 5, I argued for building platforms that can be reused multiple times, to serve as foundations for more specific systems. For platforms to be used in this way, they need to be findable in some repository. There's no point in investing in building a reusable banking platform if no one can find it.

When the code is shared by various groups in the company, there are directories, and each one of them has a file with the name of the owners of the code in that directory, the key people who need to authorize any modifications to it. The tools that work well and provide productivity gains are integrated, so that you can search (directories are huge, as there are millions and millions of lines of code in a company) across the directories to find what you want.

Here's another story that shows how this kind of integration can even prevent incidents. Brendan Burns is an engineer who worked with me at Google and then at Microsoft, where he is now a corporate vice president. When we were at Google, we came across a bug in the search system, which happened very rarely, but needed to be solved. He discovered the root cause, which was an error in the code, as if they were Lego bricks connected in such a way that, depending on the movement, could cause the whole structure to collapse (an inaccurate metaphor, but it goes with our Lego theme). Based on this diagnosis, Brendan searched Google's entire repository looking for patterns like that one and found instances in three other teams. Despite having the problem, they hadn't yet suffered any incidents, because it only manifested itself in certain very rare situations. With access to the entire Google codebase, an engineer was able to implement the fixes and prevent possible future incidents.

And the advantages don't stop there, because in addition to speeding things up, promoting collaborative and asynchronous work, and allowing faults to be corrected in series, the **right tools boost learning** and the culture of testing. When a developer finds a file in the repository that has something close to what they're looking for (a platform, or a module, which is something smaller) and reads the code, they also have access to the tests which not only validate that code, but also serve as a demonstration of its functionality. A test to validate a banking application demonstrates how those APIs are used, and the developer can read the tests and understand the system more deeply, independently from the developers who wrote the code.

On the tools side, Google's organization was very cohesive, which doesn't mean that there weren't duplicates of other kinds of systems in the company. In the early 2010s, the company was experiencing a growth boom, with a style of management very bottom-up, which helps explain how we once discovered that there were multiple teams working on personalization based on user profiles, all applying different techniques. Each one of them, however, were carrying out their explorations based on frameworks that were built upon the same basic building blocks (the same "bikes").

Duplication, if it occurred, was limited to more specific areas, such as personalization—where it was welcome, since no one knew yet what the best solution was, and we needed to experiment. Once we solved the customization problem we were working on, potentially generating the best solution for some scenario, that solution would become part of a platform for all the teams and would be used in that way from then on, reducing future duplication. Working in this way, libraries become more and more complete as new solutions are incorporated, leaving developers free to concentrate on business differentiation.

It's important to remember that Google's tools solution is not replicable, because it was conceived and built in a different era, when cloud services didn't exist. The computing framework that served as their infrastructure was built from scratch. This would no longer be the case, as today there are a multitude of outsourced tools for any company to choose from. Google itself provides many of these tools with its cloud service, Google Compute Platform (GCP).

Today's engineering tools don't have to be created in-house, but you need to choose them well and use them consistently. There are many options for programming languages, development environments, testing environments, code review systems, systems for putting code into production and performing automatic validations. For all of these, there are a variety of tools, often provided as pre-assembled packages. Technology leadership must be very intentional in how it uses these services, which need to be chosen rationally and standardized across the various areas of the company.

Other companies, such as Meta, have followed Google's lead in standardizing their tools and using a single repository for the entire company. In Brazil, an example of this is Nubank, which has taken standardization very seriously since its inception, going to the extreme of using only one programming language for the entire company—which has its pros and cons.

Easy-to-read code

Developers spend more time reading code than writing it, which is why the organization of code affects the productivity of any company. A smaller number of languages and the standardization of formats help professionals to quickly understand what they are reading, know whether they need to make changes, and find problems. Rather than imposing guidelines determining how code should be written, ideally the tools adopted by the company would automatically ensure this standardization.

There are huge arguments about where to put the parenthesis in the code, whether the content of the parenthesis goes on the same line, what to put on the next line, whether each line should have a maximum of 80 characters because that's the right length for printing or whether it's all irrelevant. When a company has a single integrated tool for everyone, formatting can be done automatically, no matter how the person has typed. No one needs to spend their energy checking if the code is formatted correctly or culturally reinforcing these guidelines, because they are simply automatic. Developers' eyes get much better focused on reading efficiently when formatting is consistent across the company. This is very important at critical moments, such as when an incident occurs.

A basic concept in programming is that everything you do with one language can be done with any other.[35] There is no language that allows you to create more possibilities, in the same way as in human language: everything you can produce in English, be it prose, poetry, or a social media post, can also be done in French or Greek. It's the same for computer languages, because Python doesn't do anything that Java can't do. Even if the result is the same, however, there are advantages to standardizing and **limiting the number of programming languages** used in an organization, as I explain in the following sections.

Fewer languages increase communality

In addition to the multiplicity of repositories and outsourced services, it's common to find companies utilizing a Tower of Babel of programming languages, sometimes with more than 20 of them in use, because of the disorderly growth of technology and the lack of direction. Developers' instinct is to design customized processes suiting their personal preferences. The situation gets even worse when the company has departments that don't talk to each other and work siloed, each one tackling projects in their own way, with no idea of how

---

35. If it is Turing-complete (https://en.wikipedia.org/wiki/ Turing_completeness). Everything that is computable, i.e., that can be solved by a computer, can be implemented in any programming language that is "complete" (Turing-complete), such as all the usual programming languages (C++, Java, C#, Go, Python, Lisp, etc.).

the engineering staff in another department are dealing with similar problems. Often there is no one dictating how the tools should be used, or which languages are preferred.

The best of all possibilities is when a company starts off already on the right path, as was the case with Google, because this avoids future technical debt and provides a perpetual advantage. When this doesn't happen, the sooner an intervention is made to focus on tools and increase standardization, the better, because as the company grows and languages and third-party services proliferate, any attempt to change things later will seem imposing and limiting for developers. The tendency is for each developer to continue working in the language and tools they like best and try to preserve the code they have already written.

Using common tools and software allows teams to be interdependent and to work together, which is positive for the company. Sometimes, collaboration must be somewhat imposed because the functionalities that need to be produced for the business are more complex than what each employee could achieve working alone. One way to encourage exchanges is to use code as collaboration. Clear code, with common languages and standards, increases the ability of teams to understand each other.

With fewer programming languages, engineers become more fungible within the organization, because it's easier to move from one team to another. If each team decides to write code in a different way and someone changes teams, newcomers must spend their time learning the new tools and standards and it will take them longer to become productive there. If a project gets delayed, key people from other teams can step in and solve the problem. Without standardized tools, all this collaboration is hampered, because whoever is going to help must first learn the standards from scratch.

The gain in agility is even more crucial in borderline situations, such as when an incident occurs or when an emergency task force is formed for a priority project. In these situations, every minute that passes without a solution can mean huge losses, and the possibility of the problem being found and resolved faster is much bigger if the company's most senior individual contributors are called in and know the code well enough to contribute.

Developing technical leaders

The standardization of tools is, in fact, essential if engineers who have progressed through the technical arm of the Y-career path[36] are to be able to exercise their leadership fully and assertively like I advocate for, working directly on the code, rather than just drawing on a whiteboard.

When an engineer reaches the status of Staff+, they understand a lot about their system already and can start working on issues involving various areas of

---

36. See Chapter 2.

the company. Their role is to define architectural standards, coming up with ideas for structuring projects, such as a single identity system for the entire company, the execution of which could be delegated to more junior developers. The Staff+ engineer is then responsible for mentoring these developers and should spend the best part of their time reviewing the code.

When someone is recruited externally, the uniformity of the tools means that they will be able to acclimatize more quickly after being hired, having the ability to intervene promptly in different areas. Someone arriving at a new company in a leadership role necessarily has a lot to learn, and a standardized engineering foundation will make this adaptation much easier.

There is more than one path for individual contributors in an organization, as discussed in the book *The Staff Engineer's Path*, which I mentioned earlier. This path can be like that of Google developer Sanjay Ghemawat—someone who picks a complex piece of the business and becomes an expert in it, spending most of their time writing and reviewing code. Another possible path is that of an architect, who gives direction, integrates teams, and makes things work. At Microsoft, for part of my career, I worked more along these lines as the architect for Azure Compute.

Without standardized engineering tools, companies tend to have a disproportionate number of architects and to relegate code to very junior employees. Google was even against the term "software architect," which is a generic name for a Staff+ professional who spends less time in front of their computer and more time drawing squares and circles on the whiteboard. Instead of architects shackled to the drawing board, they wanted to have a much larger number of senior developers, those who, even at high levels of leadership, have a day-to-day hands-on job, or rather, hands on coding. They're just much better developers technically, dealing with much more complicated code and giving directions to the more junior developers.

It is preferable for professionals to continue writing code throughout their career, even as they venture into different areas of the company. Imagine a potential code "artist," a future Sanjay, who is training in a company where the custom is for technical leadership to stay away from coding. It's quite possible that this person will think at some point: "If I pursue a career as an individual contributor, I'll be forced to stop writing code and I'll only have to draw on the whiteboard or hold team alignment meetings." Maybe they would decide to settle down and stay in the company, without developing their talent. However, it is more likely that they would look for another job where they could continue to cultivate their career as a developer. Either way, it will have been a huge loss for the company.

An organization can have so-called software architects who only do a little programming. These professionals are thinking about new technologies, focused on H3, the longer-term horizon. They have the macro vision and the temperament to conduct relevant meetings, convince the company's top management of the technology area's needs, and speak publicly on behalf of the company. As I mentioned before, I myself did this type of work as Azure Compute's architect. The point is that a company needs a very small number of professionals of this type.

At Microsoft, I held a lot of meetings using the whiteboard, and I used to joke that on my whiteboard, Azure worked perfectly, and no incidents ever happened—very differently from real life because, as I'll point out in the next chapter, incidents always happen. This kind of meeting with circles and boxes drawn on whiteboards is useful when you're doing high-level planning and defining technical direction, but there will inevitably come a time when you'll have to make the transition from the whiteboard to code. It's in code that everything happens.

Thus, the existence and daily use of standardized tools are determining factors in the effectiveness of the Y-career path in producing technical experts at the top levels of leadership who participate directly in the code and maintain its quality.

The neglected value of code reviews

I've been talking a lot about "writing code" up to now, but technical leadership is much more active on a day-to-day basis in code review. This process is often considered less important or purely bureaucratic but which I consider central to any technology company.

A large part of an organization's technical direction should be done through code review, not meetings or presentations, and for this to work efficiently, engineering tools must be used.

I've already mentioned the power of **code review as a training, coaching, and mentoring strategy** in the chapters on people in Part I, and I'll go into the subject in more detail here. To begin with, code review can be seen as a continuation of the onboarding process for newcomers joining the company. With careful and detailed code reviews, you can teach them what the processes are, how the system is organized and what details need more attention. The code review process is extremely important for mentoring newcomers, especially when working remotely.

By reviewing code attentively and carefully, the company's technical leadership professionals can convey the technical direction for building the systems that will provide stability and will enable innovation. Of course, this doesn't mean that an individual contributor at the highest levels should spend all day reviewing code for systems that are commodities and will have little strategic relevance for the company. They may even do that to gauge the quality of work at the bottom of the pyramid, but the bulk of their activities will be concentrated on the most relevant and strategic projects for the company at the time.

Care for the code generates efficiency because it's easier to correct a problem during the review and testing than after the system is already live. Obviously, some incidents are also avoided along the way if whoever is responsible for the review feels comfortable asking for modifications whenever necessary, which only happens when the whole process is ingrained in the company's culture. The review can't be seen as a task to be done in a hurry just to get the code into production straight away.

Regardless of whether a person writes text or code, mistakes are bound to happen, and the person who made the mistake usually can't see it, no matter how carefully they re-read everything. In texts, as in this book, there is a process of editing and then proofreading, and with code it's the same: a second and even a third pair of eyes greatly reduces the likelihood of errors passing unnoticed, even if several tests have been carried out. The code reviewer brings a fresh, different set of eyes that can catch a bug or suggest a modification that will make the software better.

Rules for effective code reviews

The clearer the guidelines and standards are in the company, the easier it is to review the code, because what falls outside the standard is obvious and not open to questioning. When there are no standards, the review is much more subjective. If a tool already integrates formatting, for example, the reviewer won't even need to add comments to correct the way the code is formatted. Even if the guidelines aren't integrated into the tool, just the fact that there is a standard in place removes the need to add commentary, because suffice it to say: "It doesn't follow the standard," and no further explanation is required.

There is an established etiquette that must be followed when reviewing code, because the review must be respectful, and although speed is not the main objective, agility also counts. First, it must be clear to everyone that the review is not something personal. Both the reviewer and the reviewee must treat it as constructive feedback, not a fight of egos. The person doing the code review shouldn't compete with the person writing the code. The worst thing that can happen are those "diminisher" leaders—in the terminology of Liz Wiseman and Greg McKeown in the book *Multipliers*—who receive a code review request, reject it wholesale, and then rewrite it in their own way. When a leader does this, it's not uncommon for employees to get frustrated and, over time, leave the company.

Code reviewers must have professional maturity to realize that they are not competing with whoever wrote the code, but rather looking after the quality of the result, imposing the company's architectural standards, and trying to prevent incidents from happening in the future. Both the person who asked for the review and the reviewer must keep their comments polite and objective, with the awareness that the purpose of their exchange is to improve the code so that everyone wins later.

On the more practical side, it is non-negotiable that any request for a code review should come with the corresponding tests. As soon as the reviewer receives the request, they should open it, take a quick look to check that the tests have been included, and give an initial response to whoever wrote the code, already opening a line of communication. If they see that there are no tests, or that something is not up to standard, or that for some reason they are not the right person to do the review, they should return the request immediately, so that no time is wasted, and a new submission is done correctly.

The existence of tools to conduct these iterations increases the effectiveness of the process. Comments from both sides are recorded and should be read by the other people involved in the project, because then learning is multiplied. The tone of the comments, both on the side of the reviewer and the reviewee, needs to be polite and clear. As much as I'm a fan of using standard tools for code review, if at some point the conversation gets too long, a live chat can sometimes be the best communication tool. When there's a disagreement that isn't being overcome, it's worth asking for outside help from a third party to mediate the "negotiation."

In addition to checking the correctness and unambiguousness of the code, the reviewer should have broader objectives in mind, after all, this is how they put the company's technical guidelines into practice. In the review, they assess whether the proposed change is being made in the right place, whether it really solves the problem, whether the problem is the right one to be solved and whether it is a necessary investment. In architectural terms, the review checks that the design of the change is scalable, that it ties in well with other elements of the platform, that it is resilient to failures, and that it won't create dependencies that lead to technical debts. It also checks, through the tests submitted, that the proposed interfaces are extensible and intuitive, and that there is correct documentation, because failures in these aspects can lead to maintenance problems later. It is the reviewer's responsibility to ensure that all new code has been written in the spirit of facilitating further testing in future modifications, and to analyze how resilient the code is to instabilities, and whether there are mechanisms to prevent potential incidents in production, as well as systems for alerting and reversing changes in the event of an emergency. The reviewer's considerations also extend to any performance problems that the change may cause, which is more common than you might think, so there needs to be specific tests in this regard.

When a leader does a code review, they can ascertain in practice whether the concept of platforms described in the previous chapter is being applied and should act if they find that it is not. They also ensure that the code is clear, readable, and standardized, so that any developer in the company should be able to read and understand it, even many years later. Finally, they are helping to build platform mindset in the organization to avoid technical debt; plan short-, medium-, and long-term horizons; and foster innovation.

Valuing code reviews

To establish a strong code review culture in the company, it needs to be prioritized, and this prioritization needs to be reflected in people's performance and compensation reviews. If the system only values new features that have been put into production, the importance of code review goes unnoticed and there is no incentive to do it. How can we reward a person who spends 80% of their time carefully reviewing the code written by other engineers, ensuring that the business priorities are being translated into systems?

The emphasis on code review must be aligned with the parameters used for career progression, and other less tangible initiatives can reinforce the process. First, set the example: if the company's top technical leadership reviews code and makes their comments publicly available, this sends a clear message to the rest of the engineering teams of how important code reviews are—leading by example is very effective, as described by Ben Horowitz in *What You Do Is Who You Are*, which I've cited before.

At Microsoft, Girish Bablani and I got into the habit of inviting the team's top code reviewers to dinner every month. They were selected by some criteria we established, such as number of reviews, total number of comments, and number of comments that led to changes during the review process. It was a way of bringing the engineers closer to the leaders in a more informal setting, and of giving visibility to the importance of code reviews, as we circulated a communication naming the best reviewers of the month.

Leaders' personal involvement is one mechanism for recognizing achievements and establishing company culture. Strategic objectives must be integrated into both performance reviews and more informal incentive methods.

Another reason to value a code review culture is to share responsibility when an incident occurs. The person who reviewed the code cannot be seen as the one who simply authorized the change by pressing a button, but rather as a partner in the development of that code, and someone who can get involved in solving any problems that may arise.

I'll talk more about incidents in the next chapter, but here I think it's crucial to advocate for technology companies to have a strong culture that supports code review and testing. When I arrived at Stone, the testing culture wasn't as widespread as I would have liked, and since then I've been working to reinforce it. The habit of exhaustive testing makes it possible for teams to feel more secure in opening their code to external contributions, be it from within the company or even outside it, in a truly collaborative way.

Inner source and open source

Open source codebases are those that are stored in public repositories, to which people from all over the world have access to use and contribute improvements. The owners of the repository grant approvals in a distributed and asynchronous process, with governance that is usually not from an organization but from a group of collaborators who wrote the software. The idea is based on the notion of community and has been working for many years, with many people contributing, improving, and correcting errors, as with the Linux operating system.

A helpful concept is to reproduce within the company environment a limited version of open source, which can be called inner source, enjoying all the benefits of open source software development, but keeping it protected within the organization. There are tools that support this type of workflow.

With strong test coverage and a good code review culture, there is less fear that a modification from outside the team will introduce a problem. Open source works because rigorous testing is a requirement, and review is done responsibly and thoroughly.

Engineering tools should give you the feeling that you are doing inner source, collaborative work. And, if it makes sense for any of the libraries you're building to seek external collaborations in general, it's worth considering making it open source.

At Yahoo, for example, we worked on a project to synchronize distributed components in a library called ZooKeeper, which ended up becoming open source. This was something that came naturally, and today it receives many voluntary contributions and is widely used, under the administration of The Apache Software Foundation.[37] When a locally developed system is not the core of the company, it is useful to consider opening the code. Of course, that's not something you're going to do with a system that is a competitive differentiator for the company, but there are many other systems that could benefit from mass collaboration.

Focusing on engineering tools

Someone who has been convinced by arguments like the ones I've outlined and who wants to standardize the tools used in their company will certainly face opposition. There are those who will say: "I have three engineering teams, it's better to let them choose how they want to work." One of the challenges is that there is no ideal set of tools, because for certain tasks one is always more efficient than the other. My answer to this dilemma is that it's worse for the company to be paralyzed waiting for a perfect system, because then you create a scenario where each engineering team has its own set of tools, and the whole company misses out on the benefits of the commonality provided by standardization. It's preferable to have a well-defined set of tools, even if it's not ideal for some of the teams or for some scenarios.

For an established company thinking about standardizing its tools, the first step is to create an **internal**

---

37. Available at https://zookeeper.apache.org/.

team dedicated to the engineering toolset. The role of this team is not directly linked to the company's business goals, as its main objective is to make developers more productive. In practice, about 10% of the total engineering team should be allocated to work on engineering tools. It is essential that this group is comprised of great professionals, but hiring them can be challenging due to the scarcity of people with this profile in a market that places little value on behind-the-scenes work. The team needs to be senior and prestigious enough to enforce standardization at all levels.

Depending on the size of the company, the project to standardize engineering tooling can become a long-term plan that demands heavy investment. It requires limiting the number of programming languages; choosing, building, and migrating tools and platforms; and reviewing the organization of the company's repositories. All this must happen without paralyzing the production of new code and the company's day-to-day activities.

It may be challenging to justify to the CFO of a company with 200 developers that they need to fund a new team of 20 people to define and manage the engineering tools. The simple—and truthful—justification that "collaboration will be much better in the future" probably won't stick. I'm writing this book precisely to help with the argument so that the tooling reorganization project isn't put aside just because, after all, the company needs to fund that other new feature for next month's launch.

A simple example of productivity could help convince those responsible for the money. When I was at Google, I used the time between meetings, even if only half an hour, to code, either adding a test or writing a new module, however small, because the tools facilitated these tasks. There was a single repository in the cloud and always synchronized. If the repository weren't synchronized, it would first have to be synchronized with the local computer, and then that half hour would be gone. The kind of productivity gain makes an incredible difference to the workforce.

The road to standardization

Imposing standards usually creates some resistance, so we need to promote a cultural transformation to convince people of the importance of this initiative. The standardization of programming languages is a particularly sensitive point, and I'll explain below the process I've adopted in Stone's transformation. The decision over which languages to adopt cannot be trivialized. Criteria must be used, such as the popularity of the language within the company, how easy it is to hire people who know that language well, and if good tools and libraries exist.

It is also advisable to look at the market and understand what is being done elsewhere. At Stone, we spent a year talking to people to benchmark the set of languages and tools that made sense. We talked to other players in the national and international market, with technology suppliers such as Amazon and Microsoft, we developed proofs of concept with prototypes, we tested, we saw what worked for us and what didn't work. If a solution is very efficient but no one in the market is adopting it, it is important to understand why, since it might not be worth committing to a radically original path.

On the topic of programming languages, although language choices can be encapsulated by teams—since the entire interface between systems is done through APIs—for more exotic languages there may not be as varied a computing framework as is available for other, more popular languages. Perhaps, for example, there isn't a suitable testing environment, and the company would need to create its own. A more popular language will have more tools to choose from, more open source systems, more libraries, and so on.

There is no exact recommendation on how many languages to adopt within a technology company. When I was at Google, we used C++ for 90% of things, and Android, which was a separate open source project, was written in Java. Specific systems may need special programming languages, such as embedded software, the kind that runs on Stone's credit card processing machines, for example. Since there aren't as much memory and computing resources available there, we prefer to use C++, which is a more efficient language for this kind of environment.

There's no need to take extreme stances. No company needs to have 60 languages, but neither is it necessary to adopt a single language. An intermediate alternative is to have a language for each scenario: one for everything backend, one for frontend, one for data manipulation, and one for embedded software. Limiting the number of languages makes it easier to maintain standardization.

Culture change

No matter how rational the criteria used to standardize languages and tools, those affected tend to resent an imposition where there were no restrictions before. Standardization is necessarily imposing, and that's why I prefer it to be done as early as possible, because once the restrictions have been imposed, for each new person who arrives at the company they will no longer feel like restrictions. They will just be the company standard. For those already in the company, resistance will only be eased if there is strong engineering leadership and a culture based on growth mindset.

Showing the team how much flexibility standardization allows helps in the effort to convince them. Tools enable people to work asynchronously and remotely, in different time zones, with a better quality of life. You almost always win points by swapping a meeting for asynchronous collaboration, and I'm sure that engineers like to do engineering more than going to meetings.

A technology company that fosters a platform mindset is one in which communication is efficient and done whenever possible through tools. This toolset should be standardized enough for developers to feel agile and productive to take care of the code, and be open to collaboration, in an environment that values learning, fungibility between teams, and the construction of platforms that will lead to innovation.

I deeply value people's talent, but I don't want a culture of heroes. Knowledge must be disseminated in a way that is integrated into the workflow, in mentoring, in code reviews, and even in sporadic meetings here and there. The "artists," those great talents I alluded to in previous chapters, have the role of inspiring and training other people, but they can't do everything themselves.

I witnessed many of the brilliant Andrei Broder's incredible ideas during the years I worked with him, but there were many he wouldn't have been able to put into practice hadn't he been surrounded by other professionals to help with the implementation, such as when he imagined using information retrieval techniques for advertising systems and created the area of computational advertising. On top of his original idea, others emerged, with my participation and that of Vanja Josifovski, among others, and the project was a success. Without Andrei, the initial idea wouldn't have existed, but the team around him also had a lot of merit.

In the successful case of Resource Central, the platform for predicting the use of Azure services that we built at Microsoft,[38] neither Ricardo Bianchini nor

---

38. See Chapter 5 on reusable platforms.

I can claim the credit for this achievement. Eli Cortez deserves most of the credit. He was there day and night, leading the team, managing the platform, and running it in production.

Not even Candido Portinari himself, being Candido Portinari, would have been able to execute the "War and Peace" panels without the support of tools and a group of people. And if the tools and guidelines imposed any limitations—not exceeding the space available for the murals in the UN General Assembly and following the format commissioned by the Brazilian government—this was no obstacle to the final result achieving the grandeur and importance that we all know the work has.

HOW TO USE ENGINEERING TOOLS TO PROMOTE COLLABORATION

- Use a single repository, or a set of integrated repositories, searchable with a search tool.
- Promote intentionality in the choice and standardization of engineering tools.
- Limit the number of programming languages, favoring fungibility across the company.
- Define a robust culture of testing and code review, with concrete incentives.
- Adopt collaborative inner source and open source systems.
- Have a dedicated team of experienced professionals focused on developer productivity.

Chapter 7

Incidents and post-mortems—the "horses"

If the CEO of any company announces publicly that they're going to bring in a new CTO to finally put an end to incidents, I must warn you right away that they'll be spending money on hiring for nothing, because that's such an unrealistic expectation that even the best CTO in Silicon Valley wouldn't be able to live up to it. And I'll go further: not only are incidents bound to happen, but they are also welcome to a certain extent, as I'll explain throughout this chapter. I want to have incidents! Those who never had to face incidents won't ever be prepared to deal with them when they inevitably arise. That doesn't mean, however, that you should sit back and wait for them to happen. You must work actively to prevent them and contain their consequences.

In the following pages, I will talk about how it is necessary to have teams well equipped to deal with incidents, as well as automated tools and processes to minimize them, and clear leadership guidelines so that they are, in the end, learning opportunities in the present and drivers of structuring actions in the short-, medium-, and long-term. I will discuss the importance of post-mortem meetings to understand what happened and look for the root causes of the problem. I will also address the importance of a blameless corporate culture to create a work environment in which everyone can feel safe to make mistakes (and admit to them) and can be comfortable to carry out their tasks with the responsibility and boldness needed to innovate—as I said before, I like to call this **fearless execution**.

My main point in this chapter is that incidents are to be expected. No one should be reprimanded just for having been involved in one. A good CTO, then, is the one who, applying a platform mindset, establishes a culture that values collaborative work and plans a system of technical checks and balances to deal with incidents, not the one who promises to put an end to them.

Incidents, people, and decisions

On November 11, 2022, known in many countries as Single's Day due to the numerical representation 11-11 and a famous trade holiday in China, we determined that we would have a production freeze and wouldn't make any changes to Stone's systems. We serve Chinese customers and didn't want to run any risk of incidents during a strong sales date for one of our partners. Nevertheless, close to midday, the system went down.

As usual, we quickly organized a meeting to understand the situation and to mitigate the incident as soon as possible. We were all perplexed and puzzled as to what technical reasons could have triggered the problem, especially since there was an explicit instruction not to deploy any changes into production during that period to guarantee the stability of the system. I asked everyone and heard categorical statements that no one had touched anything. I remember asking everyone in the meeting more than once: "But how did a system crash without anyone making any changes or deploying anything?" I started to think of increasingly unlikely situations, and even called the data center engineer to find out if they had done any changes on their side that could have impacted us—the answer was negative.

After several more minutes of an emergency operation involving dozens of people trying to find the source of the crash and solve it, suddenly, as if by magic, the system stabilized, and everything returned to normal.

Someone had circumvented the directive to freeze system modifications and deployed new code into production that day. Upon realizing that he had caused the instability we feared, the person didn't warn anyone, ignoring the frantic efforts of the team dealing with the incident, and simply reverted his change, bringing the system back to its previous state. This action alone could have caused another incident on top of the first one. Luckily, everything worked. The person thought the incident was solved and that was the end of that.

But sweeping things under the rug is not the way to deal with an incident, as there is always a risk that it will surface again if the engineering team does not properly understand the cause. So, we continued with the investigation and finally got to the root cause and the employee in question. He was working on a non-priority project and had still circumvented our freeze guidelines.

I was forced to fire him. I took the opportunity to write a statement to the whole company about the reasons that led me to my decision to dismiss him. I wanted to be as transparent as possible. Basically, I said that anyone has the right to make mistakes at Stone and shouldn't be afraid of being fired for it, but it's essential to have the maturity to admit your mistakes, help find a solution together with the rest of the team, and understand that, by breaking guidelines, there's a real risk of losing credibility not only with the company's leadership, but also with our clients. The fact that this employee didn't take responsibility for the error meant that several people in the technical leadership team had to dedicate themselves to investigating in the dark several other possibilities for the system crash. What's more, he was in the meeting when I asked about changes and didn't inform us of his two deployments—the original one that caused the crash and the reversal.

I follow technical incidents and I know that they are part of the nature of anyone who works with technology, as I said in the introduction to the chapter, but this particular incident went beyond a technical issue to bigger ones: repeated non-compliance with guidelines for not doing deployments and a lack of trust in the company's culture of seeking solutions to problems without pointing the finger at the culprits. If the person had said something along the lines of "Guys, I'm sorry, I did this and that," we would have had the opportunity to quickly fix the problem, minimizing the impact of the incident, and move on. The person would have been instructed not to break the rules again and we would probably have reinforced educational actions and measures to prevent episodes like this from happening in the future, without the need for dismissal. But that's not what happened. As well as not identifying himself to clarify the crash, the person made a second illicit deployment to reverse the error, risking causing another problem with the new change introduced without supervision. Making a mistake is understandable and forgivable, but making two changes without code review is very serious, and lying to everyone multiple times is even worse.

Incidents are a part of life (and technology)

I really like the book *The Black Swan: The Impact of the Highly Improbable*,[39] by Nassim Nicholas Taleb, for its idea that the unexpected is an integral part of life and we can, therefore, take advantage of it. The book's title refers to the fact that we can't assume that black swans don't exist, as has happened before in human

---

39. *The Black Swan: The Impact of the Highly Improbable*, by Nassim Nicholas Taleb (Random House, 2008).

history, just because we've never seen one. This kind of thinking is also fundamental to the analysis of computer systems. We cannot believe that there are no bugs and that the system is correct. We must assume that there are bugs and they just haven't manifested themselves in production yet.

Incidents often occur because some part of the system is tampered with, leading to a different code execution path and triggering the problem—the manifestation of that bug that had been sitting there for a while. A principle of software engineering is that if a piece of code has not been tested, it has bugs—in large-scale systems, it is impossible to test every interaction, which means that dormant bugs exist and will eventually appear.

The culture of testing is extremely important, but it's impossible to predict every situation. There isn't a single company in the world where there aren't incidents, because they are part of the game—and of life, as Nicholas Taleb preaches. You can be sure that it will always take you 10 minutes to get your child from home to school, based on the experience of the hundreds of journeys you've made on the same route, but one day it will take you 50 minutes because a tree fell in the middle of the road and stopped all traffic in the area. Events like this happen, and there's no way of predicting them.

That's also the idea behind incidents. Problems arise without anyone anticipating them. Engineering processes are designed to minimize incidents, and these processes should be rigorous enough to reduce the scale of incidents. The worst-case scenario is ignoring an incident that happened and allowing it to manifest again. This type of situation is no longer an incident, but a recurring problem that becomes a technical debt.

As an engineering leader, I don't want recurring problems, only one-off incidents, which happen from time to time. When an incident is identified, short-, medium-, and long-term structuring actions must be activated so that it doesn't happen again. It's important that we assimilate the logic of the black swan and prepare ourselves to mitigate incidents quickly when they happen, minimizing the impact on our customers.

Luiz André Barroso used to tell the story (if it's true or legend, I don't know) of a large-scale incident where one of Google's data centers became totally disconnected. When people investigated the cause of the service interruption, they discovered that it was...a horse. Somewhere near a Google data center in South America, a horse died, and they dug a deep hole to bury it. It turns out that in the process of digging, they ended up hitting the underground network cables connecting the data center to Google's network, which took it completely offline. I recounted this story of the horse during Stone Camp, Stone's leadership event that takes place every holiday season, and it had such repercussions that "horse happens" and "horses will continue to die" practically became mantras in the company.

What can we learn from this story? Could this horse incident have been avoided? Probably not, but once it was discovered that this possibility existed, it became necessary to question whether the cables shouldn't be installed even deeper, or whether they shouldn't be encased in some kind of cut-proof metal. You can't assume that Google is going to dig up all its existing data centers around the world to bury cables deeper or cover them with metal on top. How much would these structural actions cost? Millions of dollars, and it would be an endless job. If, on the other hand, the cables in the next data centers were already built to be resilient to horse deaths, the incident would have yielded a more feasible structural plan.

There are immediate actions that can be taken for incidents, for example when the cause of an incident was the lack of sufficient testing. Reinforcing the testing culture and integrating it as much as possible into the tools is a fundamental measure to avoid bugs. Imagine, in an almost cartoonish example, if such a mistake were to result in a bank deposit being understood by the system as a subtraction from someone's current account, rather than an addition of money. The person deposits 200 dollars in their bank account and realizes that they are 200 dollars poorer. It would be a terrible situation in every sense, and it's always worse when the fault in the system is noticed by the customer before it's detected by the company.

Monitoring mechanisms

Companies need automatic alerts and monitoring systems to detect incidents before they become public knowledge. Nobody wants to have to rely on customers notifying them of failures, as in the absurd hypothetical example I mentioned. Once internal alerts notify of an incident, you need a structured incident management process to contain it.

Once an alert is triggered, someone on call is activated. This person is responsible for analyzing what is happening using the organization's monitoring mechanisms. These mechanisms pinpoint which systems are having problems and why they are not working, helping us solve the situation as quickly as possible. When a system has a problem, the initial goal is to put out the fire as soon as possible, sometimes in a palliative way. A definitive solution to the problem can be a later step.

Cloud computing providers face an even more delicate situation when incidents occur, since so many people around the world depend on them. Because of this, they publish official communications and reports explaining in detail what happened in serious incidents. This is also the case when a mass-use service, like Instagram, goes down: the company usually creates a page to give users and stakeholders an explanation. This type of external communication may not be a regulatory requirement, but it is good practice, especially for companies that can have a substantial impact on their customers, such as cloud providers and financial institutions.

When a critical incident occurs, an online emergency meeting, also known as "war room," is opened to try to resolve the problem, and a specific person is assigned command. It is not always possible to know the root cause of an incident immediately after it has been resolved, because the initial goal, and the most urgent, is to mitigate the problem. Once the incident is under control, the post-mortem exercise is to investigate what happened and write a document about that incident, explaining the root cause and defining short-, medium-, and long-term actions to avoid and minimize the occurrence of similar incidents in the future.

A short-term action might be to create an alert for the situation if it recurs. A hypothetical medium-term action may be to create more instructions in the change management system so that it can't be bypassed or circumvented—this may be a significant action in the system and take months to implement. A long-term action could take years, as it could be related, for example, to resolving a major old technical debt, requiring more structured and planned actions. Ideally, companies should already have a list of technical debts that need to be monitored and solved at any given point. These outstanding issues should be considered during the planning cycle and weighed against other features that need to be developed.

A thousand alerts, zero alerts

It would be great if a company could predict all possible incident scenarios. There would be no more problems. But even with the best monitoring system in the universe, the best alert system in the world, a perfect code review system, the best developers, and a good process of automatic CI/CD (continuous integration, continuous delivery) checks, the best the engineering leadership would be able to do is to minimize incidents, but not fully eliminate them. There will always be some incidents, because the interactions between systems are complex and in constant evolution, and it's impossible to validate everything. People move in and out of a company, employees don't understand the protocols, one change interferes with another—there are many variables. Both code and people are subject to failure.

When it comes to alerts, having too many alerts is bad. Having a thousand alerts is the same as having zero alerts. What you want are alerts that are relevant, that really draw attention. For example, for cloud providers it doesn't make sense to alert every time a hard disk fails because, given the number of disks they have in all their data centers around the world and the likelihood of disk failures, alerts would go off nonstop. Rather than alerting, it would be better to invest in an automated system to deal with these failures.

Undoing a change that has caused a problem is not always easy or viable. Developers need to take this into account, and include the ability to do automatic rollback, which means undoing the change in production if problems arise.

If *The New York Times* decides to overhaul its graphic design, for instance, the newspaper's technology team won't have the chance to test the new look on every combination of device, browser, and computer system on the planet to guarantee that it will work universally. To avoid problems, developers could make the change allowing the content to be displayed with both the new and the old design. This is not trivial. Another measure is to put the new design into production, but without it enabling it. The content continues to be rendered in the previous design, with the new one in production but protected by a switch. This switch can be a flag, such as DESIGN-VERSION, which can be changed from 13 to 14, and thus change how users view the site's content. If the new version doesn't work, we can revert the value of the flag back to 13 without any major problems. This must be **thought out intentionally**, which is sometimes difficult because, depending on the type of change, it can be very hard or impossible to go back to the way things were before—there is no way to revert to credit cards without chips once the clients' physical cards have chips.

Ideally, systems have tests that validate how the system is running in production, providing health indicators that can be used to create a general health dashboard. When we realize that the health of the system has been affected by some change, we can automatically go back to the previous version and start an investigation to understand what happened.

In Microsoft's Azure there is a huge combination of types of servers, network routers, data center designs, and software versions. Imagine the disparity that exists, with more than 60 data center regions spread all over the world and a very high number of software changes per day. Over time, the heterogeneity that accumulates in a cloud server park is enormous. When developing any new version of the system, the change management process must be rigorous. When I worked at Microsoft, one thing we tried to do was to capture as much of this heterogeneity as possible when validating changes. For example, instead of deploying the change fully in the United States and then discovering that it didn't work in Brazil due to a hardware incompatibility, we tried to deploy it 5% in the US, 5% in Brazil, and so on, to slowly build up confidence that it would work globally.

Often when we had an incident, involved parties would immediately blame it on the complexity of Azure. "Why?" I'd ask. "We validated it, but we didn't test it on the XYZ combination, because Azure is too complex and has too many possible variations." My answer was always the same: the root cause of an incident is that whoever wrote the code didn't protect themselves from this complexity. If the function introduced wasn't validated simply because it was assumed that all the routers in the world would behave in the same way, it was to be expected that it would cause a problem sooner or later. **Complexity is not the problem**. Complexity is the natural state of the world, which, moreover, is heterogeneous.

I go back to the concept in the book *The Black Swan* of not being able to predict every possible situation. Obviously, you can't foresee every interaction in a system, but if every little part is tested rationally, the chances of a major problem occurring when the parts are combined are reduced. However, if no tests are implemented, the chance of an unforeseen interaction causing a problem is enormous. As I said before, if it hasn't been tested, it's bound to have bugs.

The way forward is to limit the scope of a problem and have agile mechanisms to undo any changes that cause incidents. A good practice for an application, for example, is to not roll out the new feature to everyone, but only to an initial user base of 1% of users and then get feedback to check that everything is OK. If it is, you move on to 5% of users, then 25%, and so on, until you reach everyone. It's worth crawling at first and, once you're doing well enough, you can walk and run afterwards.

The millennium bug: avoidable and unavoidable incidents

Usually, an incident occurs when there is a change. Not always—no one at Google had changed anything when the horse died and the data center was affected, according to Luiz André Barroso's story, but that's usually what happens. If a team doesn't do anything new, the system doesn't change and the chances of exposing a problem are very low. But nobody wants a technology area frozen in time. The pace of change is usually fast, so it's usual for an incident to be caused by a recent modification that hasn't been well tested, revealing a vulnerability.

It is rare that an external event will cause an incident. At the turn of the year 2000, there was widespread concern about computer incidents, because the number of digits used to represent the year what was going to change. Engineers all over the world had to anticipate possible problems in systems that assumed the beginning 19xx for years, only considering the last two digits, and make all the code changes so that the year "00" wouldn't be understood as 1900 instead of 2000. The Y2K problem, or millennium bug, as this situation came to be known, meant that an incident had to be proactively avoided. If no one had thought of this, several systems could have crashed, possibly causing serious incidents on a global scale.

Apart from these exceptional cases of external causes, foreseeable or unforeseeable, it is more common that incidents are caused by internal changes. Therefore, mitigation often involves locating the change that caused the problem and reversing it as soon as possible, returning the system to its previous state. Once in this previous state, everything should work again. This is why tests, validations, and alerts are so important. With them in place, when you deploy a change into production, automatic tests monitoring the health of the system detect faults so that a rollback can be carried out and the system can return to the way it was before the unwanted change was deployed.

When a code is approved by the code review process, it goes into a pipeline to go into production using the CI/CD (continuous integration, continuous delivery) system. This pipeline runs more tests and more validations. In the production phase, automated tests need to keep monitoring the state of things, as if it were a person who has their temperature taken every five minutes to detect a possible fever. Unlike nurses and mothers, computers don't get tired of doing these tests. These automated tests are running all the time, and once you realize that a system is in poor health, you can understand what is going on to try to prevent a major incident.

In many cases, when the problem has been triggered by a change, once you reverse it, the problem is gone. More complicated cases may involve more elaborate actions. For example, if a change has caused the credit card transaction system to go down, you can go back to the way things were before, but what about all the transactions that weren't processed in the meantime? Possibly palliative actions will have to be taken to allow people to complete their payments. It's a delicate situation, because it could be that someone has tried a few times to pay for a service by credit card and, as the operation wasn't working, decided to pay using other mechanisms. There could then be duplicate payments. If this happens to several people, the next day the company will have a huge line of people calling the customer service department to complain, which would cause a customer service incident. Situations like this can cause a **chain reaction,** such as technical problems causing related customer service problems.

Another example of a chain reaction: when someone tries to log in to the Stone application, they need to provide their username and password. If the login system is down, the user may think that the problem is on their side, and not with the system. We once had a situation where several people who couldn't log in to the application decided to uninstall it and reinstall it to see if that would make the problem go away. The issue was that, to prevent fraud, users must have registered devices. The uninstallation ended up generating the need for a new registration, and a lot of people ended up having trouble with the app, flooding our customer service team with requests. In this case, an intervention such as warning the customer that the login system was unstable could have prevented the mass uninstallation of the application and the customer service incident.

Tools, tests, and AI to combat incidents

Since we assume that incidents are intrinsic to technology, the best course of action is to create resilient systems. However, there's no such thing as perfect, which is also why it's unrealistic to spend your life just increasing test coverage for existing systems, without producing anything new.

The development conveyor belt, composed of tools, processes, and validations, should reach such a level of maturity that it guarantees the engineering and product teams the agility to innovate without worrying too much about the stability of the system, what I call fearless execution. Most platforms should already have been tested, so that the chance of a team "breaking" something big is low. The more technical debt, disparities, and lack of validations, the greater the team's insecurity.

There are certain areas of a company, such as new product development, that need to move fast and validate their ideas quickly. Well-designed platforms should insulate these areas from major problems, so that a problematic change doesn't generate a tsunami of setbacks.

Artificial intelligence tools such as Amazon Code-Whisperer and GitHub Copilot help a lot, because they can be integrated into the company's development environment and perform various functions. You can point to a piece of code you've written, and it generates unit tests that validate that code. With this support from AI, the degree of trust should increase. AI increases productivity, but humans still must get involved to monitor whether that generated code is good, and if it is, send it for code review. If the artificial intelligence hallucinates and creates something that doesn't work, the code review process should come into play. Engineering processes already involve humans naturally (human in the loop), which makes the use of co-pilots for code development and testing a very suitable AI application.

Automation against breaches

Stone's payment system is one of the most robust in the world, and there are almost never any incidents related to it. But we did have a historic incident in September 2022, and that incident was a failure due to a change made by a developer plus several other contributing factors. It was a bit like Malcolm Gladwell's[40] description of plane crashes, which according to the author, are the result of a series of small simultaneous errors that culminate in a big problem. For Gladwell, the airplane system has so many safeguards that, for a plane to crash, the pilot must be asleep, and the system must fail, and it must be raining, and the engine must break down, and so on. If all these events don't happen together, the plane won't crash. Our "plane" crashed that September due to a set of protections that all failed simultaneously.

But a mere two months after the incident, another one happened—the Single's Day system crash, which I described in the beginning of this chapter. The engineering leadership had given explicit guidelines that we would be freezing production changes on that November day to not risk any system instability or an incident. But the freeze wasn't automated in the system, it was just a general guideline for the team.

Reflecting on the Single's Day incident, it became clear afterwards that we had left room for circumvent-

---

40. Malcolm Gladwell talks about plane crashes in his book *Outliers: The Story of Success* (Penguin, 2008).

ing processes. If there was only one automated way to deploy code in production, using a tool, the manual deployment that the employee performed would simply not have occurred. The system had loopholes left by us, and we needed to better lock down our infrastructure with more effective tools to guarantee the correctness of operations from then on.

Another loophole we identified on the Single's Day incident was that the system had allowed someone to make changes on their own, circumventing our change management system. Armed with this information, we created restrictions to prevent this from happening again.

Ideally, the tools themselves should be able to categorically prohibit an unauthorized change from happening. In other words, they should allow the leadership to impose the freeze directly on the tool, without ambiguity or room for non-compliance. The conclusion of the Single's Day case from a technical point of view was that the system wasn't as tight as it should have been and required structural actions to shield it better.

You can try to avoid incidents with freezes, which are a proactive way for the leadership not to expose the system to instabilities at times that it considers sensitive for the business, as was the case on Single's Day. However, in addition to warning engineers about the freeze, it is necessary to have blocks in the system itself that reinforce the directive. The counterpoint is that imposing freezes is not the best solution for the business itself, because it is more productive to have deployments happening continuously. Blocking deployments, or only executing them within certain timeframes, means that when they finally go into production, there could be a lot of accumulated changes. What if there's a bug? With fewer accumulated changes, it's much easier to understand what happened and to mitigate the incident more quickly. That's why frequent deployments of smaller changes are ideal.

A week without deployments generates an accumulation of seven days of changes. If something goes wrong, it's extremely complex to look for causes with so many new things deployed at the same time. The reality is that the fewer deployments freezes the team can sustain, the easier it is to determine the cause of problems and minimize incidents.

Incidents during the day, when everyone is in the office or working online, are easier to deal with than at times when most people are asleep. Sometimes, for very critical situations, there is a choice to do deployments at the weekend or in the early hours of the morning, with staff on duty, but in general, deployments should happen during regular working hours. If they cause micro-incidents during these times, there are more staff present to monitor and mitigate the impact. The process of managing incidents is more natural, more fluid, and more agile with deployments taking place all the time.

Why, why, why...?

So far, I've talked about how to prevent and minimize incidents. Now I would like to take a closer look at the next step, which is identifying the root cause. The process after an incident is mitigated usually involves a meeting with the engineers involved, who try to locate the root cause of the problem. The main goal during an incident is to get the system back up and running to control the damage. Then we move on to cleaning up anything that's been left behind, investigating and identifying the root cause, and writing the post-mortem, which is a report describing what happened and proposing improvements for the future.

For the post-mortem, I like the **5 whys** strategy[41] developed at Toyota to solve quality problems with their products. The technique advocates asking "why?" several times: why did this incident happen? Is it the complexity of the system? No. Then what caused the incident? A horse died! So what? The horse died and the hole they dug to bury it cut the server cable. Ah, so the root cause is the cut cable, not the horse's death. How likely is it that this will happen again? From there, it's possible to start thinking about some structuring actions, so similar problems don't happen in data centers that we'll build in the future. The

---

41. The 5 whys investigative technique for understanding the origin of a problem was first created in the early decades of the 20th century by Japanese inventor Sakichi Toyoda and applied at his family's Toyota car manufacturer.

purpose of looking for the root cause and writing the post-mortem is to generate learning and prevent similar problems from recurring.

In the case of the Single's Day crash, the questions would be: "Why did the system crash?" Because an employee made an unauthorized change. Is that the root cause? Not yet. Why did he make the change? Because the system allowed it. Why did the system allow it? Because it allowed loopholes for manual deployment. So, a root cause is this vulnerability in our infrastructure and our processes, and that's where we need to act.

We want to make the environment increasingly safe, with improvements derived from these learnings. Another very important point is that all of this must take place in a **blameless** environment. Blameless postmortem is an analysis of incidents without pointing fingers. In principle, no one is responsible for the incident, it just happened. And the more checks and balances there are in the engineering life cycle, the more the focus is taken away from looking for culprits. It is true that, in most cases, it is a new code written by somebody that causes an incident, but the validations should have ensured that it wouldn't be a problem. Human failures are to be expected and this needs to be integrated into the risk management system.

The purpose of the post-mortem process is to improve the state of the platforms, tools, systems, and processes. The aim is to raise the bar so that incidents don't happen so often.

Leadership must constantly reinforce that the company culture is blameless, otherwise no one will believe it. I talk about this frequently with everyone in the company. I hold a weekly post-mortem forum, where we review the previous week's incidents, and the company's entire engineering team is invited to participate. The concept of blamelessness is always reinforced on these occasions: "Don't forget that we're presenting the post-mortem here, and it's for learning, not for pointing fingers." The Single's Day episode was atypical in that it involved a dismissal, precisely because the person didn't trust the company's blameless culture and didn't assume his mistakes.

A blameless culture

We talked about hiring the best people and developing good processes to retain and develop them. Part of this is also to have a culture in which employees feel comfortable that causing incidents would not have negative repercussions for their careers. If a developer does not do anything and doesn't deploy any code into production, they don't run the risk of causing incidents, but the company also doesn't produce anything. It's a worst-case scenario for all parties.

I want the team to work with the mindset of fearless execution. It's vital for a tech person to be able to work in this way, because that's what allows creativity and innovation to flourish. But to be able to act with no fear, we need to have controls in place. Anyone who writes code needs to be judicious, testing appropriately, but they should also know that everyone in the company understands there are system and human failures, and it's impossible to protect yourself from all of them.

When Stone sponsored a marketing campaign for the 2023 edition of *Big Brother Brazil*, one of the most popular shows in the country, my top priority was to make sure that if we had incidents due to a high number of accesses during the show, we would have an automatic mechanism to get back online. I prefer to assume that incidents will happen and protect myself, minimizing the mitigation time, than to hope incidents won't happen. The blameless culture revolves around this too. There will always be a human component in any incident.

In the Single's Day incident there was a mixture of human and technical problems. If the person had intended to circumvent the change management system but had been unable to do so, the incident would not have happened. If the loophole in the system had been there and the person had followed the guideline not to deploy, the incident would not have happened either. There was also the possibility that the person had written the code and shared it with other professionals for code review, which would have generated a record of the change and would have allowed the code reviewers to catch errors. With testing, the person could have found the problem with their change before putting it into production. That deployment, however, was not reviewed at all. Many errors in tandem contributed to the incident.

But in that episode, there was every chance for the leadership team, me included, to learn that the blameless culture needed to be reinforced more, and that it needed to be made clearer to employees that no one had to be a hero and do everything alone. Spreading the culture is fundamental so that people won't want to do things in secret and that they really embrace the philosophy of collaborative work.

If all this wasn't transparent on that occasion, it would have been a failure of leadership, and I take my responsibility. Doing a "post-post-mortem" here, perhaps I had assumed that the blameless culture was more ingrained in the organization than it actually was. Perhaps I wasn't as assertive in my approach to the subject as I needed to be to make people feel safe about assuming their mistakes.

Dismissals because of incidents shouldn't happen in technology companies, because making mistakes, taking responsibility, and fixing it needs to be an established culture. Incident forums tend to be tough, but I understand that we need to constantly reinforce that goal is to learn, not to blame anyone.

Teams that never lose don't know how to recover

Repeated incidents create the fear that there is some system vulnerability that will produce the same situation again. The engineering team gets paranoid about these situations, and rightly so. The engineering leader must be present and calmly set the tone to not stir things up even more. Nobody can think strategically when they're desperate. An incident war room must be objective, investigative, and collaborative. Calmness needs to prevail for the solution to emerge.

I want to highlight a crucial point: incidents are welcome. Without incidents, an engineering team doesn't develop the skills to mitigate them, doesn't learn to hold post-mortem meetings, doesn't have to rethink its own processes. Having incidents is part of our daily lives in tech, and no one can wish for zero incidents in their company. We should, however, minimize the impact on the customer when they occur.

Establishing a parallel with the world of sports, a team without incidents is like the team that wins every game in the tournament but loses the final match. To avoid panic when something goes wrong, it may even be beneficial to lose a few games along the way so that the team can identify its own weaknesses and be able to reorganize and react in a timely manner. In the same way, incidents are learning moments, which end up nurturing the culture of platform mindset.

Sometimes it's even worth holding an incident simulation just to gauge the team's reaction. It's important for the team to be trained in how to behave in an emergency. A completely new situation in which the company is losing millions of dollars a minute generates enormous pressure, and the team must know how to act professionally under pressure.

When I joined Stone, the practice of holding post-mortem meetings or looking for the root cause of incidents with a blameless spirit was not yet widespread. Over the last few years, I've been trying to set the culture. But it's hard work—building a culture in any company takes time and patience.

IMPORTANT POINTS ABOUT INCIDENTS

- Raise awareness that incidents will always happen—and should happen, because they are a sign that we have systems in production.
- Hold post-mortem meetings to find the root cause of incidents and turn the experience into a learning exercise.
- Develop a blameless culture and assertive communication of this culture, promoting a safe environment for dealing with incidents.
- Promote tranquility to manage incidents without pressure on the team.
- Strive for continuous and frequent deployments to only generate micro-incidents instead of macro-incidents.
- Develop the required skills to handle incidents, if possible, in simulations or in incidents of limited impact on customers.

PART III
LEADERSHIP

Chapter 8
Inclusive work environment

For a leader to serve as a role model, in addition to being inspiring, they need to be seen as someone who can also make mistakes and go through difficulties. So, in the name of honoring this commitment to being a more transparent human being and being true to my limitations, in 2021, I went against the conventional policy of separating personal and professional life and published a rather confessional article on LinkedIn. I'm reproducing an excerpt of this content below because of its relevance to the subject of this chapter:

Significant events and circumstances have had a profound impact on my personal life, which inadvertently impacted my professional life. These experiences existed as both tailwinds, which helped me in a positive way, and headwinds, which had negative effects on me.

By far, the biggest tailwind in my life is my family. This was true when I was growing up and still holds true today. Growing up, felt loved and protected by my parents, and I knew they would support me no matter what. They instilled in us the value of education and hard work, but above all, they believed in us so much that I started to believe in myself. This confidence enabled me to go to grad-

uate school, move to the US, switch jobs, and be confident that even if I failed in all of that, I'd still have my parents' support.

During my first year of college, my father was diagnosed with lymphoma and given a prognosis of six months to live. This was in 1991, when cancer treatments were not as advanced as they are today. Losing my father would have been a headwind that I don't know how I would have recovered from. During this time, I, subconsciously, decided that I would be the best student I could be and that I would graduate as fast as possible. My father miraculously survived seven years of intense chemotherapy and radiation treatments and has been cancer free since. What could have been the biggest headwind of my life, turned into an experience that made me focused and driven.

I got married in 1999 and in 2006, after I had been living in the US for seven years, my wife was diagnosed with breast cancer. The cancer was very aggressive and within 18 months, despite all the numerous surgeries and treatments, she passed way. Seeing someone I loved so dearly deteriorate from a healthy, active person to someone who could not climb up a set of stairs made me feel powerless in a way that was both draining and humbling. While this was obviously difficult for both of us, I learned to balance my work and take care of her, and to focus on the most important things. I learned to be empathetic towards myself when I failed, or when I was simply too tired to do anything. Through this difficult time, I developed an outlook on life that strengthened my sense of gratitude and mindfulness, which I try to utilize every day.

After being remarried for four years, and having two very young daughters, my second marriage came to an end. I didn't know how to react. In the case of my first wife, I was convinced that I did all that I could to help her fight cancer, whereas in this case, I was blaming myself for not doing enough to try to salvage the marriage. I was ashamed of myself—to the point that it took years for me to be able to talk about this to some friends and coworkers. For a long time, I was hiding this part of my life by avoiding social events and talking about my personal life at work. However, this is another example that the biggest headwinds are the ones that can shape, and possibly help you the most. My divorce helped me prioritize and balance work in a more sustainable way. By focusing on my family, I not only became a better father, and very close to my daughters, but I also learned how to have a more balanced life. The years since then have been the most fulfilling of my life, both personally and at work.

It was a huge challenge to open up so publicly on a social network, to talk about my family, to tell others about Renata's death, my father's illness, and my divorce, but I wrote and published the text because I felt I needed to show my reality, the paths that had shaped me and led me to become, at that time, technical fellow and corporate vice president of engineering at Microsoft. By talking about the adversities I went through, I wanted to reinforce, especially for those starting out in their careers, that there will always be headwinds that at first only seem to push you backwards.

Another message I tried to convey with this article was that the difficulties served to make me a leader who values an organization where everyone can be authentic in their work environment, without having to hide any part of themselves. I still believe and promote that today.

When I published this article, Microsoft was investing to raise awareness among leaders about issues of diversity and inclusion (D&I), and to create a common language so that these very relevant and inflammatory issues wouldn't end up being misunderstood. After all, it's all too common that, afraid of speaking out or doing something stupid, people decide to simply keep quiet with their arms crossed.

It is in this position that I place myself in this chapter, that of someone who does not intend to dictate perfect solutions to the fundamental mission of promoting more equitable conditions for all, and who risks addressing the subject because it is better to err than to remain silent. I will emphasize here the need for collaboration and participation by a diverse group of people at all hierarchical levels of the company because, without this, complex tasks cannot be done and innovation is hampered. I also advocate for the diversity of ideas and for psychological safety, so that everyone can express themselves without fear or repression.

I also present my thoughts on what leadership needs to do to create a respectful and inclusive work environment. While it is urgent that every company adopts active policies to reproduce the racial and gender composition of the population in its teams, it is also extremely important to go beyond just improving the hiring process. Real inclusion allows people who belong to minorities and socially disadvantaged groups to feel comfortable collaborating, being productive, growing within the organization, and becoming an example and inspiration for people inside and outside the company.

Accents and vulnerabilities

A great example of a leader who publicly shows his vulnerabilities is Satya Nadella, CEO of Microsoft. I even remember a meeting in which he showed a slide about his weaknesses. The message he wanted to convey was that you can be flawed and still reach the top of a company like Microsoft. And it was with this empathetic attitude that he spearheaded the transformation of the company's culture, as he describes in his book *Hit Refresh*.[42]

One of my vulnerabilities is that I have difficulty expressing myself in public with the eloquence I'd like. No matter how hard I try, no amount of coaching can make me give charismatic talks or interviews like Jean Paul Jacob did when he spoke on behalf of IBM at the end of the last century or deliver speeches with the clarity and fluidity of a Barack Obama.

---

42. *Hit Refresh: The Quest to Rediscover Microsoft's Soul and Imagine a Better Future for Everyone*, by Satya Nadella (Harper Business, 2017).

Over the years, however, I have relied on different types of specialized help to improve myself as a human being and as a leader. The text that introduces the chapter was encouraged by leadership coach Gia Storms, with whom I worked not only on personal development but also on the complex issues of diversity and inclusion in the workplace. Gia gave me the strength to write in a very personal tone and encouraged me to speak in public even if I was imperfect, because there is a positive side to this exposure: by realizing that the leader has weaknesses, people end up taking comfort in allowing themselves to have their own weaknesses. It would be much less painful for me to avoid this kind of situation, but I try to embrace it, preparing myself as best I can. It's not simple, and my consolation is that it might help someone who is shy about speaking publicly or needs to improve in another area.

When I joined IBM in 2000, I had only been in the United States for a year, having just finished my post-doctorate at Princeton. My English was reasonable, but still needed to improve to express more complex thoughts and ideas. Even today, after more than 20 years living in the US, I have an accent—as my daughters, "legitimate" Americans, are keen to point out. At the beginning of my career at IBM, there were situations when I wasn't allowed to present a project of mine at important meetings solely because of my "strong accent."

My experience was like that of many foreign collaborators, especially against the backdrop of the technology boom at the turn of the millennium, which attracted people from all over the world to the United States. The fact that an industry leader exposes himself publicly by speaking English with an accent can go some ways towards combating prejudice.

Members of disadvantaged groups who reach high hierarchical positions are implicitly responsible because they act as role models and inspirations. Nigerian engineer Abolade Gbadegesin, a technical fellow at Microsoft, is an example of someone who has embraced this role. He has positioned himself as an African exponent since joining the company in 1998 and has become increasingly active as he has grown in his career.

Abolade firmly believes—and I agree with him—in the importance of people in groups of interest to belong to a community, and that was his goal when he brought Microsoft's African employees together, initially mainly for social events and personal exchanges. Over time, however, the group of Africans began to realize that they could have a more direct impact, acting to combat the career obstacles that black employees generally face, and so they moved on to more structural actions. The immediate goal was to train a generation of black leaders at Microsoft through a mentoring and support program, which eventually lead to bringing a Microsoft Development Center to Africa.

The ACD (Africa Development Center) was inaugurated in 2019 and is based in Nairobi (Kenya) and Lagos (Nigeria). Its mission is to break away from a more welfare-based approach to Africa and instead bring business and development there.

In 2020, when anti-racist protests erupted in the United States following the murder of George Floyd by a white police officer, Abolade and I decided to have him talk to the team, and one of the questions was about the fear that (white) people can feel about talking about racial prejudice, fearing to reveal their unconscious biases, or of simply letting something silly slip out. He replied that the only way to approach these highly complex subjects is by talking, even if mistakes are made initially. That's what I think too: an environment of respectful conversation allows people to exchange ideas and do better over time.

Visible representation

My work with Gia Storms was important in starting to educate myself about D&I. I immersed myself in the material she gave me, in books, podcasts, texts, and interviews. But then it became clear that there was no point in having the right vocabulary without applying it in practice. Aligned with my manager at Microsoft, Girish Bablani, I defined D&I objectives and included them in my annual performance targets. Leadership agreement is essential in this type of initiative, because management must know that part of the employee's working time will be dedicated to interpersonal or diversity issues.

I chose to focus on the work environment, to make it as inclusive and welcoming as possible. At the same time, by observing metrics on the speed of promotion between different groups, I tried to determine whether there was fairness and what the critical points for improvement were.

Another of my objectives was to become more visible in the company's Latino groups, just as Abolade did with Africans. It became clear to me that it was necessary for employees in general to know that I came from Latin America, so that I could serve as a role model and strengthen Microsoft's Latino community. Although there was a contingent of Latinos in leadership positions at Microsoft, they weren't necessarily visible if they didn't have "obvious" traits. Many people thought I was Russian, German, or French.

As I've pointed out before, role models within the company are important to inspire and create a sense of belonging. I tried to use every opportunity to declare myself Latino. Other Latin corporate vice presidents at Microsoft had the same awareness and sense of responsibility.

A recent immigrant working in another country doesn't have the social connections from school or college that local employees often have. It's not uncommon for someone who has recently moved to a new country to arrive at a company without being part of a social group, and the company should strive to provide the conditions for this person to adapt.

The effort to create a welcoming and receptive environment is an essential part of my job as a leader. My goal is to put together the best engineering team possible with the most varied human material. There's no point dedicating yourself to the subject for just one or two "diversity and inclusion weeks" during the year—it must be an integral part of the leader's daily work.

In my last few years at Microsoft, I also got closer to women's groups, with whose causes I had come to identify after my divorce. At that time, I felt for the first time the challenge of being the primary caregiver for two young children and the difficulty of reconciling raising children, personal life, and work in the company. Until my divorce, I had lived in the comfortable situation of having someone else take care of the children at home, in that super-traditional division of roles that tends to give men a freedom that mothers don't even come close to having.

Women in technology

The low participation of women is a well-known shortcoming in the technology sector. At Stone, this has been a major focus of our commitment to D&I. My work in this area, as CTO, started with raising awareness: my personal goal is to talk explicitly and frequently about gender, diversity, and psychological safety issues, and this is one of the most frequent subjects of my weekly posts to the team. I opened this dialog by clarifying some concepts, showing, for example, that when we discuss diversity, we are talking about gender, racial, or socioeconomic representation, while inclusion is about everyone feeling good about their work and wanting to stay in the company. There's no point in having one without the other.

According to a 2024 study by UNESCO (United Nations Educational, Scientific, and Cultural Organization),[43] only 35% of graduates in STEM (science, technology, engineering, and mathematics) careers worldwide are women, and they occupy only a quarter of jobs in science, engineering, and information and communication technology (ICT). In the United States, a report by the National Center for Women & Information Technology (NCWIT) pointed out that, by 2023, only 27% of computer professionals would be women. In Brazil, the proportion of female STEM graduates was only 26% in 2020.[44]

Due to a series of social conventions that are still in force, women go into the exact sciences in smaller numbers. And those who do go on face discrimination in the job market, pay discrepancies, slower promotion velocity, fewer opportunities in management positions, harassment and bullying, as well as the unfair challenges of reconciling work and personal life.

---

43. *Global Education Monitoring Report 2024, Gender report: Technology on her terms* (Unesco, 2024), available at https://unesdoc.unesco.org/ark:/48223/pf0000389406 (accessed May 7, 2024).

44. According to an Insper survey based on figures from the Ministry of Education's Inep (National Institute for Educational Research), available at https://www.insper.edu.br/pt/noticias/2022/5/mulheres-seguem-em-minoria-entre-graduandos-na-area-tecnologica- (accessed Nov. 21, 2024).

One of the initiatives we have adopted at Stone to increase the representation of more diverse groups is a practice whose success I had already witnessed at Microsoft: in recruitment interviews for more senior positions, there must be at least one woman and at least one person from another underrepresented group among both the interviewees and the interviewers. Even though the stated objective of this type of high-level hiring is to choose the best person for the job, without imposing racial or gender quotas, the simple act of bringing diversity to the selection process can increase the participation, for example, of women in executive positions. At Microsoft, between 2020 and 2023, with the adoption of this procedure, the proportion of executive positions held by women rose from 24% to 29.3%.[45]

It's a slow change, and I admit that not adopting quotas for the most senior positions may not be enough to change the situation in the short term, in many people's views. However, the most structural change in building a diverse workforce should come from entry-level positions, as I detail in the next chapter.

I believe, however, that firm management action allows transformation to take place organically, with mentoring programs for women (and underrepresented minorities), and by providing support so that

---

45. "Microsoft Global Diversity & Inclusion Report 2023," page 17.

they can better reconcile their careers with pregnancy and breastfeeding. The company should also provide formal training programs to combat biases and prejudices.

Unconscious bias

In 2018, we held a year-end party at Microsoft, and Ana, an engineer who used to work in my team and is now the CEO of a tech heath care startup, was very involved and excited. She successfully organized various aspects of the celebration. When, sometime later, I decided to plan a social gathering for students, I went straight to her to ask if she would like to take part in the organization. "Of course, Marcus," she said, and we began to discuss the details. I was pleased to see that she was excited about collaborating. After talking to Ana, I left work that day and I started to drive home. Halfway there, however, something clicked on me: I realized that I might have acted in a sexist way. Had I asked her to organize the event just because she was the only female on the team? I turned around, drove back to Microsoft, and went to Ana to apologize. She clarified that she really enjoyed the role of event organizer and said I shouldn't worry, but it was very clear to me at the time how easy it is to fall into stereotypes without realizing it.

Being aware of biases, those prejudices that sometimes don't even seem like prejudices, is essential to starting to change. This is one of the first steps toward transforming the culture of any team. As a leader and a manager, I want to be open to receiving the "nudges" I deserve. If the work environment is truly respectful, this feedback will be given in the spirit of education and learning.

We all have unconscious biases, which we don't realize or which take time to become clearer. I try to be receptive to others' pointing out my unconscious biases so that I can reflect and evolve. However, this doesn't mean that I don't also need to work actively to recognize my biases and prejudices.

In trying to do justice to the subject, whether in my role as CTO, in my personal life, or in the very words I've chosen to address the topic in this book, I obviously run the risk of revealing some of these biases. But I can only say for now that, at this stage of my journey in 2024, this is what I think about how to promote a diverse and inclusive environment within a technology company. I've sometimes acted and will still act in the wrong way, so I want to keep informing myself and learning. Diversity and inclusion are important parts of the principles of platform mindset that I preach.

How to be (and not to be) an ally

I was once in a meeting at Microsoft with Girish Bablani and everyone who reported directly to him, as well as a junior engineer who had been invited to give a presentation about his project. During the meeting, I thought that some colleagues were being disrespectful to the more inexperienced engineer, blaming him for certain problems and not allowing him to give explanations or conduct his presentation. Annoyed by the interruptions, I wanted to act as an ally and took a fighting stance against my colleagues, creating an unpleasant atmosphere. In the end, my intervention only made things worse, and the engineer was left in an uncomfortable situation, probably considering the meeting to be totally dysfunctional. Neither was he able to express himself, nor did I succeed in defending him.

Although my initial intention had been good, as far as I was concerned, I soon realized that I had been mistaken. After the meeting, I apologized to the engineer who was giving the presentation, trying to comfort him. I then worked with him on improvements to certain points of his presentation and proposed that we send the revised document to the team, clarifying topics that had been contentious at the meeting. With the two peers I disagreed with at the meeting, I arranged individual meetings to acknowledge that I had misstepped in defending the engineer and lost my cool, to which they also admitted that they had raised the tone more than necessary. Regardless of whether they recognized any shortcomings in their behavior, I wanted to apologize anyway.

I learned from one of my mentors that it doesn't make sense, when you want to reach an understanding, to think that you only need to admit 50% of the responsibility, giving the other person the task of admitting the other 50%. You must assume 100% of the blame, be 100% sincere and 100% willing to change, and this attitude will make the other party mobilize 100% on their side as well. That's the mindset I adopted in this episode. I went to talk to my peers admitting that I was 100% wrong.

I've come to understand that to be someone's ally, you shouldn't simply assume that they want to be "defended." This attitude can be seen as offensive and overbearing because it makes the other person look weaker than they are. During that meeting, I could, for example, have sent a private message to the engineer, asking if he would like me to intervene on his behalf, tried to show support for him through visual signals, or waited to talk to him afterward, or intervened in a neutral and restrained way. Being an ally doesn't mean becoming the hero of the situation but rather being willing to collaborate alongside someone else on their terms.

In that case, I identified with the engineer who was in the spotlight and acted rashly to defend him without thinking about the best strategy for doing so. You might think it's easy to be an ally: all you need to do is take a stance when you see something wrong because omission would already be a stance. But you must be aware that, more than a good intention, you must assess the right way to act on someone's behalf.

To improve myself, I have relied for years on coaches, and I actively seek advice. I put a lot of pressure on myself because I've been on the road for over 20 years and hold a management position. In the example of this meeting, I was much more obliged to know how to behave than the less experienced engineer. I need to communicate as effectively as possible, not just with colleagues, but also to resolve the issue for the company, which is what is at stake. If I offend someone and they leave, I create a bigger problem.

It's important for companies to provide training on how to be an ally, as Microsoft did while I was there. The material, prepared by experts in the field, explained to all employees that being an ally means being aware of your own privilege, and understanding power relations and which behaviors are not inclusive. No matter how much empathy you have, it's not easy to acquire this awareness, so commitment and a willingness to learn over a long period of time are the key points to being a true ally.

Good training shows that before acting without thinking, as I did in the meeting, you need to question your own motivations, be sufficiently informed, and ask permission from the person you want to help. This material can also include simulations of scenarios like those encountered in everyday work.

By offering training, leadership provides employees with mechanisms to help solve day-to-day problems at work. The quality of relationships acts as a lubricant for collaboration within the company. Communication without friction allows teams to work together, creating a positive environment that ultimately fosters efficiency and paves the way for innovation.

Fighting aggression

Many technology companies have been formed by young people in a highly competitive climate, with all efforts focused on growing fast and conquering the market. This has sometimes resulted in toxic work environments.

One story I heard reflects this aggressive behavior: in an interview, a female candidate was asked, "Do you prefer Lana del Rey or Nirvana?" and the interviewer then declared, "I'm glad you answered Nirvana, because if you had said Lana del Rey, I wouldn't have hired you." If they'd asked me that question, I'd have said: "It is none of your business, and I don't think I want to work for this company." If I had to answer, I would have been rejected because my answer would undoubtedly have been Lana del Rey, especially because my oldest daughter is her biggest fan. The point is that this question probably wouldn't have been asked of me, a white male candidate who didn't need to prove himself.

This stereotype is reflected in a company's culture and needs to be actively deconstructed. It's another expression of the problem of wanting to win the race by running fast, without building bicycles that would help you win future contests. It is detrimental to build a homogeneous team of Nirvana worshippers with the idea that this represents bold behavior. A company is a microcosm of society, and it can only win in the long term from a plurality of points of view and professional and personal experiences (and even musical tastes).

At Stone, we are still early in creating an environment that supports inclusion and tolerance. I can't say that my actions have turned the tide, but I believe I'm working every day to change the situation.

A few years ago, while studying inclusion and the workplace, I came across a TED Talk by Professor Christine Porath[46] on how respectful treatment between company employees is good for business. In the video, she explains how aggressive attitudes of leaders and peers cause real damage to people's productivity. The experiments she and her collaborators have carried out show that witnessing disrespectful behavior, or being exposed to rudeness, leads to errors and a drop in performance, in all types of industry.

Her work, described in detail in the book *Mastering Civility*,[47] is in line with what I've always believed. Microaggressions interfere with the productive climate and prevent the development of the receptive and creative environment that ultimately makes a company successful. I organized workshops with her for my teams at Microsoft.

Through research and controlled experiments, Porath demonstrates how bullying can quickly spread throughout a team. An aggressive leader makes their

---

46. Available at https://www.ted.com/talks/christine_porath_
    why_being_respectful_to_your_coworkers_is_good_for_
    business (accessed May 2, 2024).

47. *Mastering Civility: A Manifesto for the Workplace*, by Christine Porath (Balance, 2016).

direct reports act aggressively too, because they tend to imitate the leader's behavior—and this way of acting becomes part of the company's culture.

Christine Porath's data proves that a tense atmosphere makes people more prone to mistakes. For example, we are five times more likely to miss important information if stressed. Lack of respect also impacts commitment to the company. After being the target of aggression, 60% of people are more likely to reduce their efforts, and 12% are more likely to resign.

It's easy to see that a situation of incivility lowers the morale of the whole team, not just the person targeted by the disrespectful comment. Everyone becomes less enthusiastic about their own work.

Since a lack of civility is so negative, why does it still happen? The most worrying thing is that some leaders still think that being aggressive and rude is desirable behavior and that they need to be that way to generate fear and obedience. The data, however, shows the opposite. Employees who act with respect and politeness are more respected, are twice as likely to be recognized as leaders, and perform better in their roles.

Once, at Microsoft, a group of about 20 very senior people were working on the same technical issue, but there were many divergent paths to follow, and no conclusion could be reached. We decided to organize an all-day workshop to finally arrive at a solution. Prepared to find tempers flaring due to the issue's controversial nature, I decided to open the meeting with a statement: "We're all colleagues here. Before you open your mouth, please listen to what the other person is saying. Stop, think, internalize, and when you do speak, make sure you're adding value to the discussion and not generating more conflict." This five-minute speech at the opening of the meeting worked, because the meeting turned out to be very productive, and years later, colleagues still remembered my opening words.

Communication parameters

Some simple tools can be used for more effective communication, for those who are trying to act as allies, and for those who want to position themselves in an assertive but respectful way. One of them is the ADEAR framework, which comes from "be a dear."

- A - **Affirm** that you have nothing against the person and reinforce the value of your relationship with them.
- D - **Describe** the behavior without judgment.
- E - **Explain** what you felt.
- A - **Assume** a positive intention.
- R - **Request** or suggest a different behavior.

When you are the target of a microaggression, this framework allows you to take control of the situation and talk to the aggressor about the problem. For example:

"When we talked about budgets earlier today, you said: 'You're making drama.' Statements like that imply that I'm being overly emotional because I'm a woman and that my concerns aren't real. It makes me feel belittled. I'm assuming that's not your intention. I'd appreciate if you didn't make comments like this anymore. Is that all right?"

These are the steps: assume the person had the right intention and didn't mean to harm you, make it clear to the other person that you were hurt and explain why, and then ask for a change in behavior.

If people understand that this is a viable path, at least they have mental parameters to deal with a situation like this in their daily work. Having gone through this training, the junior engineer in the meeting I told you about could come up to me and say: "Gee, Marcus, I know you meant well, but you acted very aggressively towards your peers, and the result hurt me. I'd like you to explain what happened, from your point of view, and request that you change your behavior in the future." So, armed with these tools, he and I could have a mature conversation about that incident.

This kind of conversation is never easy, but having a roadmap of how to act can give you a good idea of where to start. Individual problems can take on a larger scale without communication parameters and quickly become institutional problems.

It is also essential not to absolve the company of responsibility for inappropriate behavior by its leaders, since we know that, in some organizations, there is still a strong culture of protecting leaders, often classified as eccentric or genius, rather than abusive and destructive.

Leadership also acts directly on the results of the culture change when it values collaborative work over individual achievements in its people's review processes. Specific metrics that show how much a person has promoted collaboration make it possible to associate concrete incentives with the cultural transformation effort.

Individual and cultural differences

Hostility can manifest in microaggressions or "little jokes." When I'm attending a lecture and someone utters a swear word, for example, or uses a metaphor with a vulgar expression, it interferes with my attention, perhaps because of a personal sensitivity or inherited upbringing.

The point is that people's cultural backgrounds are very varied, and even within the same country there are many cultural differences. Clear communication techniques reduce possible misunderstandings, which sometimes lead to conflict and resentment. The companies I worked for before Stone were all large (IBM, Yahoo, Google, and Microsoft), with employees from all over the world, including Brazil, which is heterogeneous, with many subcultures.

It's very difficult to generalize about a culture. Not everyone in Brazil likes soccer, and there are disorganized Germans who are late for meetings. With that caveat, Erin Meyer shows, in her book *The Culture*

Map,[48] that there are some important cultural patterns and differences. For example, in communication, some cultures tend to be very detail-oriented, while others focus more on the big picture. For decisions, some cultures tend to be more hierarchical, while others are more egalitarian.

The greatest contribution of Meyer's book is to show that people are different, made up of a combination of cultural and individual attributes, and that we need to listen better and ask more questions before we assume anything. Direct communication is the best way to avoid falling into the trap of understanding something that may be very different from what the other person meant to say, or of thinking that we are transmitting one message when in fact it is quite another that is being received. I prefer to think that each person I interact with is unique and deserves special attention.

Of course, I often fail and misinterpret a situation. We all have our blind spots. I even like to write these situations down (when I can identify them), as if it were a post-mortem of a "communication incident." Like any post-mortem, the focus is on learning: we can always improve.

Openness to all ideas

Imagine that a teacher presents a problem to a

___

48. *The Culture Map: Breaking Through the Invisible Boundaries of Global Business*, by Erin Meyer (PublicAffairs, 2014).

student, and the student comes up with the solution overnight. The next day, excited, he or she takes the solution to the teacher, who, to his astonishment, refuses to even look at it, saying: "There's no way you could have solved that overnight." It's perfectly plausible that the solution was correct. This kind of arrogance rules out ideas that may have great potential, and even risks stifling the creativity of a talented student.

The work environment needs to allow employees to feel comfortable expressing their needs—such as accommodation due to personal circumstances—and to share innovative ideas. If, when Vanja Josifovski said he wanted to process XML documents using a streaming algorithm, as described in Chapter 4, he had been greeted at IBM with a "You're crazy, that'll never work, we have other priorities here, and you're not going to waste your time on that," the innovation wouldn't have happened. The company would have lost the millions of dollars that were generated from that project.

Calling an idea stupid is the most effective way to kill creativity. If everyone only had brilliant ideas, our planet would be different. I welcome all ideas with open arms because there must be ten silly ones for every revolutionary idea. Suppressing ideas that still need to be improved is the same as suppressing the source of all ideas. When I receive an idea, my first move is to ask for it to be validated against real data, and this step alone is enough to act as a filter. It's a way of discussing the idea more concretely without belittling it.

I want an environment that is open to ideas that break away from the conventional, and there's nothing better for that than a team of diverse personalities who don't fit into the rules. The best projects I've worked on were born from "out of the box" ideas that I proposed and weren't dismissed out of hand just because they seemed strange. I once thought there was a better solution to an advertising serving problem that we had in production. I spent a long time thinking about it—one day, I thought it would work. The next, I was convinced there was no way out. Andrei Broder also described his thought process for coming up with a solution for detecting duplicates on the web in this way. It's part of staying in this state of conviction for a long time until the idea crystallizes. In my case, I had a safe environment in which to present an idea, even if it was half-formed so that I could collaborate with people more experienced than me to crystallize and improve it. The idea finally worked and ended up being one of the most innovative algorithms I helped design.

This creative process can be even more productive for the company if all kinds of prejudices are combated: gender, origin, racial, social, and age prejudices. Without that, very young and inexperienced people, as I was at the start of my career, can find it difficult to have innovative ideas taken seriously. Where I was, at IBM, there was also the "aggravating factor" of having real luminaries by my side, people who had, for instance, worked on the first relational database. It was natural for me to feel insecure about contributing my ideas.

However, I was fortunate to be able to express what I had thought and to see my solution accepted because the environment at IBM was receptive. In addition, my confidence came from the fact that I was brought up under the positive stimuli of my parents, especially my mother. She encouraged me to go ahead even with the chance of making mistakes. My parents probably predicted that I wouldn't do very well on my volleyball adventure, for example, as I didn't, due to my lack of motor skills. But they didn't stop supporting me when I got excited, and they were thrilled with my small advances.

Their presence in my life was and is a privilege and an advantage that, unfortunately, not everyone has. Because not everyone is lucky enough to grow up surrounded by affection and with basic needs met, leaders need to be even more intentional about fostering trust and giving support to their employees.

Accommodating diversity

As I talk a lot about diversity in the weekly posts I write for my team, I was delighted when some people got in touch with me to talk about the fact that they are on the autistic spectrum. We exchanged ideas about what could be done to make the work environment better for them, for example, with measures to limit noise or to adjust communication so that instructions were clearer. I liked that they felt comfortable coming to me and negotiating adaptations, which can be very relevant for face-to-face work. We also talked about fully adopting remote work, which since the COVID-19 pandemic, has become universal and an interesting alternative for people of this profile.

It's important that autistic or neurodivergent people in general have the chance to speak up because this allows the leadership to have a broader view of the different needs, difficulties, and potential of each team member. Everyone in the company needs to have a growth mindset, maintain a tolerant attitude, and be willing to listen to try to understand even those who may seem "different" at first.

As I progressed in my career, I realized how much unused potential there is in teams, because not all professionals have the chance to contribute to their fullest. As an introvert, it took a long time for me to start feeling at ease at my first job at IBM. Today, I don't want anyone in my team to feel that to prove themselves, they need to adopt the strategy I used back then: working millions of hours more than necessary to prove myself. It's not uncommon for employees to question what they must do to prove their worth. I emphasize to the managers in my team that it is their duty to clarify individual expectations of work, behavior, and performance. They must also think about what the company can do to unlock each person's potential, especially for people from minority groups.

Psychological safety

Once, during a meeting to discuss career progression, my team and I were evaluating individual promotion cases. As always, the discussion was very heated, and we spent considerable time analyzing each case. When it came time to evaluate an engineer I worked closely with, I spoke highly of him and supported the promotion. To my surprise, some people reacted negatively. One of them suggested that I was being unfairly favorable because of my involvement with this individual's project. I felt frustrated and was emphatic enough to make my point heard: the promotion should be unquestionable!

It didn't take long after the meeting to realize that I had overstepped the mark. The more assertive I was, the more people felt I wasn't listening to them. I behaved like a "know-it-all," rather than a leader who was open to feedback and willing to be questioned and proved wrong. This alarmed me, as I actively strive to be receptive. I took the opportunity to reflect on what triggered my behavior and why I reacted so aggressively in that situation. I concluded that, somehow, I felt threatened.

Integrity is a fundamental character trait of mine, which defines me as a person, leader, and teammate. I had a misconception that my integrity was being questioned in that promotion discussion, as if my entire identity was being challenged. As a result, I wanted to defend myself, instead of adopting a stance based on curiosity and a growth mindset.

Over the years, I've seen people feel threatened in various ways. For example, being told, "You can question anything about me, but don't question my work ethic!" This response occurs when the character trait is a significant part of a person's identity. We don't mind so much when criticism or feedback doesn't directly challenge our identity, which means I won't be offended if someone points out that I don't dress well, since being fashionable isn't a strong part of my identity.

How can people protect themselves from reacting defensively or aggressively in these situations? Awareness of the core values that make up our identity is a great starting point for recognizing what we may perceive as threats. It's a constant exercise in self-discovery, questioning, and reflection, because our identities are evolving.

Upon reflection, I realized that in the meeting I described, I could have acted differently. I could have tried to explain my perspective in the context of the proposed task, which was a promotion. If I had said, "I know I'm involved in this project, but I'm trying to be objective, and I'm going to focus on the technical challenges that may not be apparent to you," the conversation would have been more transparent and productive. Acknowledging the situation and respecting diverse perspectives, even when they differ from your own, creates a more inclusive environment. That was my first lesson.

How can we create a safe space? Threats are not always physical or easily observable. They can be verbal, implicit, and unintentional. To promote inclusion, we must create an environment where people can freely express themselves. This kind of environment naturally reduces threatening situations. Furthermore, when we reaffirm people's identity qualities, we promote psychological safety. In my situation, I was lucky enough to have a second promotion meeting already scheduled for the same week. I began that meeting by apologizing for my behavior. I was explicit about my reasons for being irritated at the previous meeting. The team and I used the experience as a learning opportunity to revisit the general principles of the promotion process in our organization and ensure that we were all aligned in the future. Ultimately, it was an episode that helped build more trust among my team.

When people feel psychologically safe, they can be more authentic, and this is beneficial for many reasons, including opening space for creativity and facilitating communication between teams, which is so fundamental in an area full of interdependencies such as technology.

It's much easier said than done, however, because it's not trivial to anticipate every situation that could be threatening. Moments that make people feel questioned are not uncommon. My second lesson was that when I believe people are feeling insecure for some reason, I pause and try to clarify the situation rather than just defend or impose a position. Finally, I learned from the episode that we won't always get it right the first time. Learning is continuous.

Another relevant factor for psychological safety within the company is encouraging employees to group according to their interests or identities. Promoting communities doesn't mean putting people into boxes: the LGBTQ+ community, the women's community, the autistic community, and so on, because one doesn't exclude the other. People are diverse in their identities and can belong to several communities simultaneously.

Community

Professor Christine Porath has recently written another book on the subject of communities. In *Mastering Community*,[49] she offers a unique perspective on the importance of building and sustaining strong, engaged communities of people, something that resonates deeply with our work as technologists.

In the world of technology, we often focus entirely on technical aspects, neglecting the power and impact of the groups around us. *Mastering Community* reminds us that we are not just building software but also participating in vibrant communities within our teams or professional networks. Through practical examples and insights that come from the world of sports, Porath explores how we can strengthen our group conviviality, promote a culture of diversity and inclusion and cultivate meaningful relationships inside and outside the world of technology. She highlights the importance of

---

49. *Mastering Community: The Surprising Ways Coming Together Moves Us from Surviving to Thriving*, by Christine Porath (Balance, 2022).

empathy, collaboration, and mutual respect in creating environments where everyone can thrive and contribute fully.

This is even more important in the reality of remote work. People should be able to work from wherever they prefer. On the other hand, I worry about junior engineers who work in a very isolated way, deprived of social interaction and the feeling of belonging to a company community.

As software engineers, we face complex challenges that require collaborative solutions. Communities are sources of support, learning, and professional growth, and should be encouraged within the company. By making room for them, the company is investing not only in its success but also in the success and resilience of the wider technology community.

I'll repeat here what I said in the chapters about technology: every impactful innovation that we build is too complex for one person to do alone. Collaboration is essential for the company to flourish, and it's much richer when we foster diversity and inclusion.

Perhaps it's too great an ambition or even a certain naivety. Still, my desire is to create such a positive work environment that the technology team feels, as Anderson Nielson says, like in the song from the renowned Brazilian artist Dominguinhos,[50] which describes a party that is so fantastic that no one wants

---

50. "Isso aqui tá bom demais," popular song by Brazilian songwriter Dominguinhos.

to leave: "Look, this is good, this is good indeed, look, those who are outside want to come in, but those who are inside won't leave."

HOW TO CREATE AN INCLUSIVE WORK ENVIRONMENT

- Show vulnerability.
- Be a role model for your communities, especially for leaders belonging to under-represented groups.
- Require the mandatory presence of women and under-represented groups in interviews, both among candidates and interviewers.
- Provide training on how to be an ally and on unconscious biases.
- Welcome diverse ideas and tolerate error.
- Promote non-aggressive personal interactions.
- Be respectful of cultural differences.
- Promote psychological safety.
- Encourage the formation of communities.

Chapter 9
Social responsibility

Why do I believe that the work of a technology leader is socially relevant? Because we can promote an internal transformation within the company that ends up impacting far beyond it. Private enterprise, in general, has the chance to be an agent of significant change, but technology companies even more so, as they have enormous potential to train people with the skills needed to improve the world we live in. These employees can then go on to train others, even in other companies, multiplying learning, economic development, and cutting-edge knowledge.

If I can convey to the teams the idea of a platform mindset, of building reusable platforms and the right tools, and of reducing technical debt and incidents, we will be able to provide better and cheaper services, which will, in turn, impact our customers and generate a virtuous cycle.

That's what this chapter is all about, where I extend the concept of a respectful work environment and employee well-being to the company's external borders. We need to be aware that everything we do reflects positively or negatively on the nearby and not-so-nearby communities in which we operate as well as society. Thus, social responsibility cannot simply be an obligation to fill a few pages of management and sustainability reports. It must be incorporated into the culture—the common good is as much of a priority as business profit.

This requires active leadership involvement in projects that make a difference, such as investing in hiring policies that look at the company's social role. Here again, I bring a personal story to show how a leader sometimes reaches a moment in their career choice that can increase the scale of their professional impact. Having grown up amidst the great contrasts of Rio de Janeiro has led me to place social responsibility as a matter of the utmost importance in my life's journey, inseparable from how I act as a coworker and think about the role of a leader.

Brazil on the agenda: Time to give back

In 2022, after more than twenty years living on the West Coast of the United States, I wanted to be closer to my sister on the other side of the country, so I organized a move to Florida. By then, my parents were already living near me and liked the idea of moving to a warmer climate. The logistics of working online away from Microsoft's headquarters and moving with everyone were challenging, as it also involved an agreement with my daughters' mother.

In this context, a few months after the decision to move to Florida, Stone approached me: the company had mapped Brazilians who had worked in big techs with a solid impact history, and my name was on their radar. They wanted someone who spoke Portuguese and had experience with technology companies to lead the transformation they were planning for Stone, with technology as the central focus of the business.

In April 2022, I took time off from Microsoft and went to Brazil for a week to do interviews and get to know the people at Stone, both the leadership and the staff. During that week, I also visited a few Stone's customers in downtown Rio de Janeiro, to understand how they used our products. I made the visits with a technology colleague and the Stone sales agent responsible for customer relations, and I was able to get a clear picture of how both customers and sales agents used Stone's technology.

What's more, I was able to witness first-hand the impact that the COVID-19 pandemic was having on Brazil. Of course, the pandemic was also profoundly affecting life and business in the United States, but since my whole family was there, I hadn't been back to Brazil since before COVID-19, and I had no idea what had happened in the country where I was born and raised. The vibrant Rio de Janeiro downtown that I remember, full of people working in the buildings and walking back and forth on its busy streets, especially at lunchtime, had disappeared—it was completely deserted. Half the places we tried to go to were closed. The scene was bleak.

"I'm in. We can decide on the salary and other details later," I said to the people at Stone, just after seeing what I'd seen. I had been telling myself for years that I wanted to help Brazil somehow, and suddenly, the opportunity came to me. "It's now or never. I won't get another chance like this," I thought. When I left Brazil in 1999, I believed I would never be able to work with cutting-edge technology in the country, so I planned to spend a year abroad perfecting my studies before returning to become a university professor. But opportunities arose, and that one year stretched into two decades.

I saw two determining factors in Stone's offer: the first was that it was a company that had a more direct impact on the end customer, with technology in the hands of users from classes C, D, E, and F, small and medium-sized entrepreneurs who needed the company's services to sell their products. Another relevant point was that Stone already had 2,000 employees in the technology area at the time, meaning that I would have the chance to directly impact 2,000 people, plus their families. I saw the multiplier potential. I knew that, one day, these employees would leave the company and start spreading the culture they had incorporated, the principles they had learned, spreading my philosophy of how to manage technology teams in a more intentional and organized way, fostering innovation.

My decision was so impulsive that no one around me could believe it, almost not even myself. I didn't even know what Stone was before they approached me. "Your brother is leaving Microsoft," my mother announced to my sister on the video call we made to tell her the news. My sister asked where I was going, and when she heard the name of a company, all she said was: "You're kidding, right?"

Microsoft tried to retain me in a few very persuasive ways, and I was very touched by the proposals made, but none of them would allow me to have the impact on Brazil that I wanted. I loved every aspect of my job at Microsoft: the company, my colleagues, my managers and mentors, and all the technology we built together. I still decided to leave, because I trusted my intuition that I could help build something amazing—a Brazilian big tech to solve local problems, helping Brazil's entrepreneurs thrive.

The recruitment process at Stone included interviews with both the company's leadership and the people who would be my direct reports. This was an aspect that I liked: the decision wasn't imposed on the team, and Stone had established that if there wasn't unanimity among the group that would report to me, I wouldn't be hired. I liked the people and was interested in the company culture and their philosophy of obsessing about the customer. I identified with the company's mission and saw that it wasn't just a slogan, that there was a clear goal of doing the right thing for our customers. I also realized that the company needed organizing and restructuring and that I could contribute to this transformation.

Joining Stone wouldn't have been feasible if I'd still been living in Seattle, since the time zone difference and the physical distance, which make traveling very long and tiring, would have made me turn down the offer. So, my first decision, to move to Florida, made the second one possible. Each of these two decisions was very big in itself, and a lot could go wrong, from moving a family with school-age children to the other side of the country to leaving a job of eight and a half years that fulfilled me.

And the process was challenging for all of us at home. We moved to Florida in June, a week after I joined Stone, and I traveled to Brazil to get to know more people at work. There was a lot of back and forth between Florida and Brazil during this period, while my daughters and parents were adjusting to life in a new city. If I hadn't had a great support system in my family, and a lot of understanding from them, things wouldn't have worked out the way they did.

The stories we tell

I've always told myself that I'd like to collaborate with Brazilian companies and invest on the country. But one of my mentors once explained to me the concept of a "front operation," which are those stories we tell ourselves repeatedly, but somehow, we never act on them. Perhaps that declared intention to help Brazil was my "front operation." How many times have I thought, "One day I'm going to do something for Brazil, I'm going to invest there, and so on," and it never happened. The opportunity to go to Stone challenged this. It allowed me to act on this dream, which came with a series of consequences, the main one being a lot of traveling and reduced time with my family.

It was a challenging transition from one company to another, from one culture to another, from one country to another. But one of the main reasons I decided to take on the role at Stone was to positively impact the team, and that's been happening. Once the transformation impacts people, they will make better systems for customers and, more importantly, they will also become better managers. I see my work as educating and training people, and Stone has made that possible.

I grew up in a middle-class family and managed to attend university, which by Brazilian standards, is a much better starting point than the average person. My family highly valued education, so I didn't have to work from an early age, and could focus on my studies. During college, a professor, Luiz Fernando Gomes Soares,[51] introduced me to a social project he and other professors were working on in an underprivileged community in Vargem Grande, a neighborhood on the west side of Rio de Janeiro. I joined the proj-

---

51. Luiz Fernando Gomes Soares (1954–2015) was a full professor in the Department of Informatics at PUC-Rio, director of the university's Telemedia Laboratory and winner of the Association of Brazilian Information Technology Companies (Assespro) award for personality of the year in 2008. His biography is available at http:// nucleodememoria.vrac.puc-rio.br/perfil/saudade/luiz-fernando-gomes-soares-1954-2015.

ect's group of volunteers. We donated some computers to the community so we could go there every week to teach programming to the local students. I did this for about two years with some other undergraduate colleagues and Professor Edward Hermann,[52] a Computer Science professor at PUC-Rio. When we started teaching, we realized that their basic math foundation was so poor (even that of their teachers, by the way) that we also started teaching math to the class. It was a small project on a small scale, but it planted something important in me about how voluntary work can have a huge impact on the lives of those it reaches. Years later, I learned that some of the project's students had been the first from that community to go to university.

Luiz Fernando organized a social project that changed the trajectory of those students. It's one thing to talk, but another to do. He knew how to mobilize the group of students at PUC-Rio and structure a program so that it was effective. I fondly remember the Tuesdays I drove with Hermann to that community to teach. It was a very rewarding period of my life and a fundamental part of my development as a human being.

There is a deficit in the training of technology professionals both in Brazil and in the rest of Latin America, so there is a great opportunity for growth

---

52. Professor Edward Hermann Haeusler coordinates the post-graduate program in the Department of Computer Science at PUC-Rio and his area of research is related to computability and computer models.

ahead. This is made even easier because the time zone is relatively close to that of the United States, which favors partnerships and projects with North American companies. Nowadays, with the possibility of remote work, many big techs see Brazil as a place that generates great professionals for their teams.

Considering that the starting salary for a computer professional is higher than for many other careers, it sometimes ends up being several times higher than the earnings of any other individual from a poorer community. Training a person even for an entry-level position in technology already impacts their lives, their families, and beyond.

Sowing talent for growth

With the increased development of tools and platforms, programming has become easier and more accessible. Imagine how complicated it used to be: you had to manually program using punching cards, which encoded the instructions to be processed by the computers. If there was an error, you had to redo the cards and wait in line for another chance to run your algorithm printed on the new cards. Today, programming languages are high-level, which means they are closer to natural language. This means that developers don't require as much understanding of computer architecture and can write code in a variety of languages with extensive libraries.

The barrier to being able to make computer systems is lower, especially when you work for a company that invests in platforms, as I explained in Chapter 5. It's not that the person will immediately be responsible for creating Bing's new search algorithm, but they could be the one to help build a website or a new application, even with a relatively low level of technical knowledge. Paying attention to this person's training from the outset is important, as is looking at someone who isn't even in the technology field but could become a developer, thus obtaining better professional opportunities and contributing to the team with diverse ways of thinking. Stone's Code[S] program seeks to attract people from other areas and train them to work as developers.

PUC-Rio also has a long-standing program with Rocinha, one of the largest slums in South America, to train young apprentices called NEAM (Núcleo de Estudo e Ação Mundo da Juventude).[53] Through this program, these young people take courses at PUC-Rio and often end up being absorbed by the university itself, working in some entry-level jobs. By becoming PUC-Rio employees, they can attend university there with full tuition scholarships.

Not everyone will end up working in a big tech company in Brazil or the United States, but some of these people will be brilliant and have the potential

---

53. NEAM began in 1981 with the training of school teachers and today has various projects, workshops and partnerships to train and educate low-income young people. More information at https://www.neam.puc-rio.br/sobre-nos/missao-e-objetivos/

to grow to have beautiful careers. The initial chance is the lever that enables talented individuals to work on gradually more complex projects as they become more able to face them.

Understanding the possibilities of a manager

The opportunity to expand my area of activity came in my final years at Microsoft, when, encouraged by Girish Bablani, I felt ready and decided to switch from an individual contributor to a manager role and led a team of over a thousand people. It was also the first time I could put my leadership and management concepts into practice with a larger number of people involved. I created a work environment that was not only stimulating and enjoyable for those who were there but also produced a series of relevant results for Microsoft. I had a great relationship with my team and all the employees, so leaving was a very difficult decision. I cried at several public farewell meetings, and the final meeting with the team was very emotional. I had even managed to get through most of the meeting, but at the very end, when I heard a colleague say, "Marcus, we love you!" I fell apart.

Part of my pride as a manager at the time concerned the socially responsible projects I was leading. As the person responsible for the efficiency of the cloud systems, I built an agenda that I liked to call "democratizing computing," since one of the objectives was to make the service cheaper and more accessible to more customers. Democratizing computing can enable people who previously didn't have a digital presence to start having one. Not only have we made the service cheaper with the efficiencies we've created, but we've also made it more environmentally sustainable. The cloud itself is a polluting system, because of the energy consumed by the servers, so my team's work in power management ended up contributing to a more positive environmental impact.

I had the chance at Microsoft to be very close to the foundation created by Bill and Melinda Gates. I wasn't directly involved in the foundation's projects, but as it was based in Seattle, I followed its projects closely and understood Bill Gates's focus on using technology to impact the world—whether it was researching a vaccine for malaria or building toilets and sanitation networks in Africa or India.

Bill Gates inspires me because he wants to solve the planet's most difficult problems with technology and scalable solutions. Even though I didn't take part in these projects, I internalized his and the foundation's idea that technology can be a tool for changing realities.

At Microsoft, I hope to have contributed to the growth of several people, as in the case of Eli Cortez, who was very junior when we started building Resource Central and is now one of the people responsible for it, as I mentioned in Chapter 5. I still mentor some of the people I worked with there, including people who have gone on to other companies or even founded their own. Because I had more career experience, I believe I helped remove some of the blockers from their paths and made the journey less arduous. Even when I wasn't formally a manager, I always valued and was involved in mentoring people, sometimes in more strategic objectives, sometimes in technical problems.

I hope that my attitude in general will also inspire people to think about leadership in a different way from the profile that is traditionally associated with people in high positions. There's still the impression that only alpha males, who are aggressive, rise through the company and reach the top. My story is not quite like that.

When I was promoted to technical fellow at Microsoft, the highest technical point in the career ladder, I received a few messages from people saying, "Wow, Marcus, how nice that someone like you was promoted." I didn't know how to interpret those messages, but I think they meant that they saw me as a more collaborative person who tried to solve things through dialog, building relationships, and changing the culture to create a better work environment rather than shouting in meetings and trying to impose my point of view.

Impacts beyond those anticipated

Technology is a vehicle for social advancement, and technology professionals have many options regarding which projects they work on and what kind of personal impact they would like to make.

Google is an ethical company that strives to produce the best for its employees and customers. When I was there, I worked on developing the YouTube algorithms that recommend the next video to watch based on the video someone is currently watching on the channel, an idea that seemed super useful and positive to us at the time. However, this is a part of the algorithm that many people nowadays think contributes to the passivity of the person who sits in front of a screen all day watching videos. In other words, an algorithm that I helped develop to improve people's experience can be perceived negatively by some viewpoints.

This is inevitable, however, as almost everything in life has a positive and a negative side. Google's search engine can be used to search for symptoms of a heart attack and save a life, or it can be used to find out how to make a homemade bomb and cause terrible damage.

When you work on basic research, you can only predict some of the ways in which it will be used in the future. The concept of algorithms I helped develop for recommending videos or ads ended up being used by several social media platforms, which can have a very varied effect on people's lives.

For example, the spread of fake news is something that worries me a lot. Suppose there has been a gain in the last decade of information reaching people faster, of someone knowing almost instantly that a threatening storm is approaching. In that case, the time needed to check information accurately has decreased, and it is no longer possible to differentiate true news from partial or totally fabricated news. People need help to filter out who is publishing what and understand whether a story is from *The New York Times* or from some business that appeared on TikTok, which shouldn't be anyone's source of journalism. I can understand the nostalgic position of reading newspapers on paper.

Another impact of technology that has two sides is the issue of online communities. There are communities for all tastes and inclinations, ranging from vintage citrus fruit growers to people preparing coups d'état or cyberattacks. In the past, those who believed in a flat earth were isolated, but now they belong to a giant community of people sharing theories and fueling social instability. Society hasn't yet adapted to the democratization of information because the technology to validate the vast amounts of data being produced in real-time still doesn't exist.

And that's why I emphasize the importance of a technology leader discerning which projects to get involved in. As I mentioned in Chapter 7, in complex systems, it is rarely possible to know in advance all the present and future component interactions or to detect something that will remain latent until it causes great damage. Climate change is an example of this: the technological progress of today's civilization is taking a toll that nobody took much into account back then when they were deforesting the planet to build buildings, cities, industries, pastures, and roads. And even with the alarm bells of science ringing in recent generations, there is no shortage of leaders shying away from taking a more active and responsible role in embracing some urgent technological choices that will help mitigate the problem.

Principled leadership

It may be impossible for leaders to understand all the future uses of an innovation developed in their company, but there are almost always more obvious risks that can be avoided. There may be alternatives that could, for example, reduce the cost of logistics but are not worth the human cost. In Stone's case, letting the people who deliver credit card processing machines to customers travel along more dangerous routes to get to customers more quickly is a risk we don't want to take. We don't want to know that we've had accidents in our logistics to generate savings, large or small.

At the time of the COVID-19 pandemic at Microsoft, we took great care to plan what health protocols would have to be implemented to protect the employees operating the data centers, since they had to be physically there.

I believe in principled leadership, a form of social responsibility. Given your limited understanding of something that neither you nor others know much about, the sensible thing is to go for a technical analysis of what could be negative consequences, and simply not go down that road if there are signs of a relevant risk.

It's always possible to reconcile ethics with the company's need to make a profit. The whole computational advertising business that we developed at Yahoo, described in Chapter 4, had strived to serve more relevant ads to the user, while still being a welcome innovation from an economic point of view.

Encouraging people to think of others

Social responsibility requires a global reflection on what you do and how technology can be used more broadly. It's not enough to think, "I'm doing a good job managing my team and delivering results." You must go further: Of all the projects your team is working on, are there any that positively impact the community?

At Stone, we are working to introduce financial education to our customers. Our goal is that, for instance, before making a commitment to take a loan, the person understands what interest rates are and what other charges may be involved. As a technology leader, I have a way of influencing the creation of this kind of program. I believe that having informed customers is a win-win situation: with more knowledge, they will engage better with our solutions.

After joining Stone, I became even more impressed with the company as I got to know its social investment initiatives better. I discovered they supported a school in the broader Sao Paulo area, the Alpha Lumen Institute, which caters to students with high abilities in all fields. Unlike the US, no schools in Brazil have gifted programs, apart from the Alpha Lumen Institute. I have been there to visit a few times and sponsor three of their students. I accompany their development closely, with regular meetings in which we talk about school, family, and their hobbies and passions. Talking to one of them, I discovered that she really liked Legos but had never owned a Lego, which was entirely within my power to fulfill.

Microsoft also had a well-structured program for leaders to embrace various causes and encourage employee participation. Everyone who joined the company in the first year received a 50-dollar philanthropy credit, which could be allocated to any non-profit of their choice, and Microsoft matched the amount, doubling the amount of the donation received by the institution. In addition, every year, Microsoft would match (up to a certain ceiling) the donations of all employees. There was even a yearly friendly championship between teams to see which raised the most money, including events and auctions. One year, my youngest daughter and I participated in a charity auction by offering a TikTok dance lesson to encourage bids, which were then donated to an institution we supported.

Microsoft could perfectly well donate a hundred dollars on its own on behalf of an employee, but the idea of giving the person the opportunity to actively think about the issue and press a button to donate was intended to encourage reflection and a culture of social responsibility.

And it's not just philanthropy that generates social impact. I conceived and led a Stone project to create an innovation laboratory within PUC-Rio, StoneLab, which seeks to strengthen collaboration between the company and the university and help create a community of exchange with a leading educational institution. The intention is to foster basic research and to create a collaboration program that other companies can replicate. PUC-Rio's objectives and mission are also linked to social impact in their local communities, and Stone chose this partnership with this in mind. One of StoneLab's projects, in the field of materials science, is investigating how to make environments cooler, which can be applied in areas of warmer climates without access to air conditioning and cooling systems. This has no direct connection to Stone's business, but by supporting it, we are sponsoring the training of scientifically minded individuals who will eventually contribute to society in other ways. There is research into quantum physics, differential equations, and others that have nothing to do with financial systems.

On the other hand, StoneLab has currently 11 supported projects with an average of 4 people per project, so if we hire one or two exceptional people to work at Stone, the investment in the whole lab will have more than paid for itself.

Social responsibility comes from the understanding, on the part of both the leader and the company, that we participate in the life of a region and must establish relationships with it. The focus must be on long-lasting relationships, not solving this or that specific problem for the company.

The Estudar Foundation[54] provides scholarships for low-income students who want to pursue graduate degrees in the United States and other parts of the world. I got very involved with them, helping select students and mentoring some of them. I also speak at several external events so that my presence can inspire other leaders to talk more about technology and how technology can be applied to improve life in the country. I like participating in events and engaging with students because I believe these can have a positive impact in people's lives.

My goal in the book is not to advertise the companies I've worked for but rather to show how engineering leaders can help influence the culture of the company they work for. I believe that anyone who works in technology should seek to engage with and contribute in some innovative way to positive social change. My decision to leave Microsoft and work for a Brazilian company is an example of this.

---

54. The Estudar Foundation was founded in 1991 and supports young people with international scholarships and leadership training. For more information, check out https://www.estudar.org.br/. All my proceeds from the Portuguese version of this book are directly donated to the Estudar Foundation, while the proceeds from this English version are distributed to organizations supported by Microsoft Philanthropies.

Promoting economic growth

Philanthropic projects positively impact the country's economy and improve society. When, for instance, Stone supports a project like that of the Alpha Lumen Institute or the Estudar Foundation, it is helping bright students who will be able to add tremendous value to companies in Brazil and the rest of the world, raising the level of technology professionals in the country. This helps everyone, including Stone itself, who will have a large pool of qualified professionals to hire from and to do business with.

Looking internally, when I talk about managing careers as if they were marathons, it's because I want someone who joins a company with a good work environment to stay there long enough to grow professionally. Not everyone will become a great developer in their first year of work. That's why it's important for technology leaders to take systematic steps to develop people—this involves many of the topics we covered so far, from hiring to developing platforms to D&I.

Innovation in both products and platforms is also a way of lowering the cost of the service and passing on the advantage to the customers. Stone provides financial services for small and medium businesses, who are not well served by traditional banks. How is it able to provide high-quality products at low costs for that market segment? By creating scalable solutions based on reusable platforms and maintaining a well-structured technology team that is productive and innovative.

I know how difficult it is for someone who owns a restaurant to optimize a digital ad campaign because they would need to understand online advertising. However, a technology leader can develop a platform to solve the problem at scale. Most businesses should soon go digital, and it is essential to train people to work on solutions for them. Thinking technology-first, a company can promote the digital transformation of its customers, boosting their competitiveness and helping to reduce social inequalities, regardless of the area in which it operates.

IMPORTANT POINTS ABOUT SOCIAL RESPONSIBILITY

- Create programs to form new technology professionals, especially from lower income communities.
- Participate in philanthropic projects with positive social impact, involving the team whenever possible.
- Choose projects wisely to maximize the positive impacts on society.
- Think thoroughly about new innovations and their possible negative implications in the future.
- Participation in external events to inspire social action by other leaders.
- Use technology to create social good, support digital transformation, and create economic value.

Chapter 10

Leadership attributes

It had been a few months since I had accepted the offer to join Stone, and I was very excited about the challenge of helping transform the company. However, the new job made my routine very hectic. While my two teenage daughters and my parents were still adjusting to their new home and all the life changes that come with moving across the country, I went back and forth from the United States to Brazil to get to know every aspect of the company in detail.

As part of the large changes that I needed to make, I ended up being consumed by work and, on many days, working from early morning until past 10 p.m. On top of all the things I had to learn about my position and the financial services industry, I was also restructuring the team, which led to countless conflicts to resolve. To make matters worse, there were a lot of tech-related incidents.

I remember one Saturday when, after a long two-week stay in Brazil, I couldn't wait to finally spend a whole day together with my family. But in the car, leaving the airport early on my way home, I was notified about an incident at work. When I got home, I barely said hello to my daughters and holed up in my home office with my computer, believing that perhaps the resolution would be quick. I had promised to take them to lunch at a restaurant, and they were waiting for me at the door, thinking they would have their father's company at any minute. The problem, however, was complicated, and I was only done at the end of the afternoon when the day was over. No one had the energy to do anything else. What's worse, I didn't fulfill my commitment to them. That affected me deeply, and I realized I couldn't continue working at that pace. I wasn't a good role model for my team or a good father. Was that what I wanted to convey to the people at Stone?

Leaders are role models. Everything I've discussed so far should be part of the leader's platform mindset: building an excellent team, fostering innovation, building reusable platforms, adopting tools, learning from incidents, encouraging diversity and inclusion, and having social responsibility. Now, I want to go a little further and reflect on the leaders themselves.

To this end, I will discuss growth mindset, the need for self-knowledge and self-criticism, and the difficulties of negotiating with the company's top management. I will also discuss the importance of sound technical judgment when choosing projects and return to engineering culture and how it can create a positive work environment.

Since I don't think there are any ready-made success formulas, what I'm going to present over the next few pages are leadership traits that I've learned from people who have inspired me and that were transformative in my quest to become a better leader.

Work-life balance

"That's not why I came to Stone," was my conclusion after that Saturday when I couldn't be with my daughters. I felt exhausted and began to reflect on what was happening. I needed to change the way I acted at work. The purpose of hiring me had been precisely to help Stone to get more organized, but what happened was that, in those initial months, I ended up assimilating the modus operandi that I wanted to change. The people there still seemed to work with the startup mentality, taking on several tasks simultaneously, no matter what, because they believed the company's survival depended on it—which was no longer true. Stone had already 15,000 employees, and an organization of this size needs to be able to function without counting on specific individuals, including the CTO.

After realizing that I had to change my attitude first, I told all my direct reports in a meeting that we couldn't keep working at that pace and that we needed to rethink routines that seemed untouchable. To begin with, people had to feel more comfortable doing something as basic as taking a vacation without feeling guilty.

In the spirit that no one's head can withstand pressure all the time, I think of what I learned from one of my mentors at Microsoft, Gaurav Sareen, now global corporate vice president of experiences and platforms. He argues that you need to schedule some breaks at work, because some emergency will always happen. There's no point in filling your schedule and then feeling unproductive because you have to drop commitments due to an unforeseen event. With breaks integrated into the agenda, there is more flexibility to accommodate whatever comes unexpectedly, without losing the peace of mind needed to make complicated decisions in times of uncertainty and strong pressure. A leader's emotional stability is contagious—as is desperation.

It's an attribute of leadership to be a role model, to demonstrate what you preach and not to keep saying, "I want people to spend more time with their families" if you don't give yourself the right to rest. When a leader works at a frantic pace, it's inevitable that employees will do the same, since it's behavior that counts, not words.

I identified the people who reported to me who seemed to be the most distressed and who were still behaving with a startup mentality, and I began to do individual work with them. "Who are the people in your team who can take over when you're not there? Do you have a backup plan for all areas? How is the on-call duty structured in case of incidents?" With this type of questioning and planning, we have been dismantling the attitude and culture of urgency at work, which often leads to overloading entire teams and burnout.

My goal was to make the following phrase true: "Apart from incidents, everything we are doing today should be able to postpone until tomorrow." Our goal was to structure the technology area so that there were no fires—real or imagined—to be dealt with every day.

A huge part of this restructuring was creating Stone's career path and improving the people review process. Employees often overwork themselves because they don't know what is expected of them. A formal career trail defines objectively what is expected of the employee at each point in their career so that the constant tension of "guessing what the manager wants" is minimized.

In a company the size of Stone, with thousands of employees, if someone feels they must solve everything themselves, it's a sign that something is very wrong and we could even have a systemic problem. As a leader, one of my most significant duties is to ensure that the company functions independently of the individual actions of the parties, including my own.

While I was restructuring the team and guiding those who reported to me to organize their own teams, I also took the opportunity to evaluate people, to understand if they could effectively lead their areas. Another initiative at the time was to hire more experienced people, following the seniorization philosophy I described in Chapter 3. Over time, with the addition of new "artists" and with mentoring, the team improved and got used to working with a more long-term outlook.

Today, two years after I took over as Stone's CTO, my personal and professional life is better organized.

My daughters have grown up and have a more independent life, and I count on the help of my parents, who live nearby. I work hard, but I can be a very present father and give the family the attention I want. I no longer go to sleep with my mind 100% occupied with work problems, with the feeling that I'm not coping with everything I must do. Changing the team's mindset, starting with my own, enabled us to become more productive and reduced anxiety and burnout.

Self-knowledge and self-criticism

I have a bedtime ritual that includes updating two lists—I love lists! The first is the list of things to do, which on a good day, has no less than ten items. Strangely enough, a very long list doesn't worry me. On the contrary, it calms me down, because I know that at some point, I'll get to the task at the end. The second list is a kind of diary, where I write down what I did well and what I didn't do well that day. These are items written down in no more than one line each, such as "I got distracted at the meeting" or "I got nervous with my parents." I always consult the list of things to do, but I hardly ever re-read this journal, and I don't even need to: the items that are repeated catch my eye when I'm writing them down, and it's easy to realize what I need to do better. If I've written several times that I got distracted in a meeting because I was answering emails, I realize I need to stop repeating the same mistake. The list helps me detect if there are patterns of behavior that I need to rethink.

The idea of making this list came about after I learned about growth mindset. I learned the term "growth mindset" when I took part in the culture transformation at Microsoft, which was based on the concept described in the book *Mindset: The New Psychology of Success* by Carol Dweck. But I must admit that, as the son of a psychoanalyst, the habit of questioning myself and trying to identify blind spots has always been very present in me.[55] The growth mindset concept proposes that people can have a fixed mindset or a growth mindset. According to Dweck, those with a fixed mindset believe that they are born with certain qualities and defects and that there is not much they can do to develop their talents and abilities beyond what is defined by their genes and temperament. In a fixed mindset, any failure is a demerit that must be avoided at all costs. On the other hand, those with a growth mindset see intelligence and talent as primarily the result of hard work and perseverance and see mistakes as opportunities to learn and grow.

By advocating that leaders expose their vulnerable side, starting by setting the example himself, Satya Nadella, CEO of Microsoft, has stimulated the vision

---

55. Regarding blind spots, I'm also guided by the principles described in the book *Leadership and Self-Deception: Getting Out of the Box*, by the Arbinger Institute (3rd edition, Berrett-Koehler Publishers, 2018), which explains how our self-centered behaviors put us in a kind of "box," with a defensive attitude that prevents us from seeing people with empathy.

in the company that failures can be an instrument for improvement, not an obstacle, and that it is always possible to learn and grow.

Satya often says that he sees three essential characteristics in leaders. The first is that the leader can see their way out of complex situations and describe in a few words what is essential and what needs to be done. Eric Boyd, who was one of my managers when I was working at Bing, was like that. We'd barely start to explain a problem, and he'd get it right away—the kind of person that's hard to find in the market, an inspiring personality who can listen to all sides, assimilate information, differentiate between what everyone is doing, and determine the right direction with inclusive decisions and simple explanations. Eric, now corporate vice president of Azure's artificial intelligence platform, is undoubtedly a model for me in this leadership attribute.

The second characteristic, according to Satya, is that those who lead must generate good energy, not in the esoteric sense, but in extracting from the team the potential it must produce, very much in the spirit of the multiplier leaders described by Liz Wiseman and Greg McKeown in the book of the same name. It's a facet that reminds me of Mark Russinovich, Azure's CTO, and Andrei Broder, both role models for their organizations, admired by everyone who works with them.

The third characteristic is that good leaders deliver results. They make things happen, promote innovation, balance short- and long-term projects, and inspire. Satya is an excellent example of this leadership trait.

It seems obvious, but it doesn't hurt to reiterate: a leader can only be good at some things, especially as they are often involved in decision-making areas that are quite different from those in which they usually have more technical experience and familiarity. Adopting a growth mindset is precisely recognizing weaknesses, knowing which characteristics you can improve on. In these areas, it is more prudent to ask for help and build teams around you that complement your skills.

Compensating for deficiencies

Being aware of one's limitations, whether in terms of technical knowledge (hard skills) or interpersonal attributes (soft skills), paves the way for building a team to help compensate for them. I know that I'm not very effective when it comes to operational processes, so I always try to work with someone who can help me with budgeting, resource allocation, planning, and dependency management.

Another practical example: when I arrived at Stone, I saw that the company had two apps, one for small and medium-sized entrepreneurs (the Stone app) and one for freelancers and micro-entrepreneurs (the Ton app). These two apps had many similar functions, such as banking and payments, but for historical reasons had been created completely independently of each other, with separate technologies. When we mapped them out, we found that the two apps had around 70% of their screens in common, with three teams working on them (because Stone's app had been developed in native code, with separate teams for the Android and iOS systems, and Ton's app was multiplatform, developed in React Native), with different levels of testing and maturity. Maintaining a structure like this is wasteful, because the more complex the apps become, the more duplicate work there is.

I then made the technical decision to unify these teams, within the general direction I set for the company. However, I had never worked directly in mobile application development, and I felt I needed to be more competent to dictate the details of this unification. I explained within the company why structurally having two distinct apps didn't make sense, and I was honest about my inexperience in the area. I decided to hire a consulting company to assess the situation, and over time, I began to look for someone to take charge of the project. The person I hired, after a lot of research, was Juliana Chahoud, who today heads up all the engineering of Stone's user experience area.

I have a solid understanding of computer science that allows me to reason about the various areas of technology, but it would be arrogant to think that this equates to mastering everything in such a vast field. If I had to carry out a project on quantum computing or cryptography, I would also have to rely on specialists to support me. Another alternative would be to stop what I'm doing and study to delve deeper into these subjects, but then I'd have to refocus my priorities and perhaps drop some important ones. I could even be damaging the company, by not dedicating my time to all my important commitments and by, perhaps, delivering work of a lower quality than that of a specialist with extensive experience in the subject could have done.

The manager's manager

It is part of the technology leader's professional commitment to learn from the CEO and the other company leaders about the other areas of the business, such as legal, human resources, and sales, and to educate them about technology so that together they can define the metrics used to evaluate the performance of the CTO and the area. In my current case, where Stone's CEO Pedro Zinner has never managed a technology company before, part of my job is to explain how to measure engineering productivity, plus everything else I talked about in this book: platforms, tools, team structure, why it is essential to leave some fat to make room for innovation, how to think about CapEx and OpEx in tech, giving him inputs so that he can manage me effectively. Another part is expanding my horizons of knowledge to the other areas of the company, focusing on how technology can integrate better with them.

Defining the innovation horizons for the company will help the technology leadership to map out investments for product innovation. If the company's strategy involves geographical expansion, for example, the CTO helps determine what technology will be needed, and over what horizon, so the strategy is actionable. In addition to strategic thinking, there is also tactical work on H1 and day-to-day operations, including system improvements, handling incidents, and reducing technical debt.

As I had never worked in a fintech before, I had to learn about specific aspects, such as the regulatory aspects of financial operations. Any change of area requires a certain amount of time for the professional to learn about the new domain—this had already happened to me within Microsoft itself when I switched from Bing to Azure.

The biggest difficulty is usually aligning expectations and deliveries, and the way to solve this is through frequent conversations and exchanging information on both sides. This is my message to those who hire technology leaders and are from outside the area: try to understand how some of the concepts described in this book work, from hiring and managing the team to platforms, tools, and incidents. It's very important for the CEO to familiarize himself or herself with these concepts and be on the same wavelength as the technology leaders.

I say this knowing that any CEO has legitimate asks coming from all sides and is therefore left with the thankless job of having to prioritize and balance these requests with limited resources. It is quite common for the technology leader to hear "no" to a request, not always because the CEO wants to say "no," but because of contingencies, such as a freeze on vacancies due to a tight budget.

There are many frustrations that cross the CTO's path: at more critical times, more speculative projects may be put on hold or canceled; regulations may change; unexpected problems may arise with suppliers. More than dealing with their own frustration, the technology leader must manage the team's anxiety and maintain morale and motivation, especially when projects are canceled, or promotions are postponed. I've experienced this in many companies.

During the COVID-19 pandemic, I was working in Azure, and the demand to run the Teams application grew enormously, with the sudden need for millions of people to work remotely due to social distancing. We needed to put projects on hold and reallocate staff to create a task force that would ensure we had the capacity in Azure to support all the important applications at that time. It wasn't just about Teams, where the work meetings were held for all the people working remotely, but also about guaranteeing the stability and availability of Azure for the emergency personnel acting on the front line, such as doctors, hospitals, and health authorities. It was an extraordinary situation that required managing the expectations of those who had to be switched from their original projects.

When making more extreme decisions, such as unifying or eliminating certain projects, the leadership should outline the guidelines for taking that course of action. It is recommended that the process is inclusive, ensuring that all the interested parties are heard, and that the verdict comes without much delay. There's no escaping it: senior leaders must deal with difficult choices, which leave a trail of disagreement and for which the best remedy is a culture of "disagree and commit" so that everyone can move on. You cannot please everyone, and delaying making hard decisions can be very detrimental.

The best way to be fair in these moments is to base the decision on a well-founded technical judgment, whether it's in favor of a platform or a tool, a team reorganization, or a change in strategic direction for a project. These costly decisions from a financial and personnel viewpoint require firm leadership—with the hope that the leaders are right most of the time.

In this respect, the technology leader needs to be technically up to date. This applies to any field, not just engineering. If a professional hasn't been programming for ten years, they shouldn't decide about tools and platforms based on the parameters they had a decade ago. For example, there are many options for cross-platform frontend development, container management, and observability tools today. These areas are constantly evolving, and engineering leaders must keep up with them to make assertive decisions.

Mission and satisfaction

Leaders must understand their mission, their team's mission, and how these missions connect to the mission of the other teams and the company. This understanding contributes to job satisfaction because the value of the work becomes clearer to all parties.

One of the pieces of advice I often give Stone newcomers in Ask Me Anything sessions is to try to explain what they do on a day-to-day basis to a layperson outside the company. An employee often joins the company and starts coding frantically without understanding the broader context of why they are doing it and how it adds value to the company. Understanding this helps create engagement and leads people to work in a way aligned with the company's objectives. For example, by thinking in terms of platforms, adopting standardized tools, embracing blameless culture, and adopting growth mindset to become multiplier leaders.

"Are you happy with what you're doing? How do you imagine your career expanding? How can you improve your impact?" These are some of the questions a mentor can ask their mentees to help them increase their job satisfaction. Subverting John F. Kennedy's famous phrase, it's not a question of asking what the employee can do for the company, but what the company can do for them. And the company won't do it out of benevolence, since the ultimate beneficiary will be the company, which will improve its productivity. An employee who feels that they are being sucked dry by the company, without receiving anything in return, will turn their back and leave as soon as they get an opportunity. As a leader, I want people to stay for the long term. Employees who stay and want to be developed become better employees in the long run.

When I arrived at Stone, many people tried to win races "by running fast." My job was to ask: "What bike would you like to build for the next competitions? What platforms can we build to make your races more efficient and less tiring?" The work often involved transferring employees to different teams to give them a fresh start, and in some cases, eliminating people who weren't aligned with the principles I was bringing in.

Obviously, it's a gradual process, and things won't change overnight, the transformation is still underway. As we slow down the team, bring in more senior professionals, focus on platforms and reusable components, and focus on testing and code reviews, it won't even be possible to adopt that unbridled rush behavior in the future. My mission is to contain productive anxiety, to show people that they may need to wait for a certain platform to be built before moving forward with a structuring project, to spread the message of intentional transformation, with planned dependencies and long-term thinking.

Programming the organization

Throughout my career, I've held various positions in technology: I was a researcher, a developer, and an architect. Later, still at Microsoft, I had the opportunity to build my team and lead a large organization, and now at Stone I'm getting another chance to do that again.

It's a position with high potential to play the role of a multiplier. I want a technology team with 1,000 people to produce as if it had 5,000, not 500. If the leader is too controlling, trying to be involved in every decision, this directly affects how the team functions. The goal is to create a learning culture, keep the team motivated, and adopt platforms and tools to boost the team's performance.

I want my team's processes to work almost like an algorithm, with no room for comments like: "That person was only promoted because they report to manager A or manager B." Processes must be consistent throughout the organization. With the definition of processes such as performance evaluation, budget management, and project planning, if you change the CTO, nothing changes, because the systems are already in place. It's practically like "meta-management," building and implementing management mechanisms that can be replicated and applied systematically.

Leaders must worry about the day-to-day operations, but at the same time, they need to think about how to "program" the organization. It's as if they were writing a computer program to automate the CTO's work, with pre-programmed solutions for most situations. Over time, as different situations occur, there will be more inputs to refine this large program. Every problem the leader faces and every new situation they encounter serves as information for changing or refining some rules so that the organization works better and better.

Another way of looking at this is to compare it to traffic control on the streets: the rules are so well defined that the driver internalizes them and just observes the signs without constant reminders explaining that you can't overtake other cars when there is a continuous line painted on the ground. The technology organization ideally works like this. These "mechanisms" (the name used at Amazon, as explained in *Working Backwards*) are defined in such a way as to make it easy to follow them. A stipulation such as "everyone in the company must use shared repositories" can be expressed in spoken or written form, or better yet, it can be automated in a tool so that even if you try, you can't go against it.

Imagine a law stating that a car can only drive at a maximum of 10 mph near a school. The simple law is a protocol that is easy to break. You could devise a mechanical solution, such as installing speed bumpers, or a more automated one, such as an electronic radar, that would make its enforcement better. But by using technology, and considering connected and autonomous cars, the vehicle itself could reduce its speed, without the user having the chance to circumvent the law. As we discussed regarding engineering tools, automation works better than written protocols. However, companies are constantly evolving, and management should be also careful to adapt to changes and prevent automation from overriding differences and making everything too impersonal.

As CTO, I decide which activities to get involved with based on their multiplier potential. Between a project that will impact a thousand developers and another that will involve ten, I prefer to work on the one that will affect more people. That's the main evaluation criteria. But throughout my career, my concept of where I could have the most impact has evolved. At first, I thought I could generate more impact by writing code and developing innovative algorithms. I also worked as an architect, and in that role, my time went to define projects and design complex systems. Now, as CTO, I have a more strategic influence, because I can define how technology can align with the company's strategic goals.

In my conversations with new hires, I often joke that I want to finish my career as a developer at Stone so that I can work with our amazing team, platforms, and tools and enjoy the company culture that I helped create.

On the other hand, there will always be more to learn as a CTO and as a human being. I want to continue broadening my understanding in many areas, and I have a lot of knowledge to gain from the young people entering the job market. Every year, I get a better sense of how they approach life's priorities differently. Technology is always evolving, so there are constant opportunities to improve platforms. The work is never finished. But I also take great pride in building teams, mentoring artists, designing bikes that will help us with future races, giving technical direction to the company, and being ready for the horses that cross my path.

For fun and profit

I'll never forget the poster that hung on the wall of our search infrastructure team's work area at Google when I was there. It said: "Searching the web for fun and profit."[56] It conveyed exactly what was happening there and the joy I felt from working in an environment with well-rewarded employees, using the best platforms available at the time, and working on very exciting new technologies. Most importantly, people worked with satisfaction, while at the same time making a huge positive impact on the company. And it wasn't just at Google that I experienced this. Throughout my career I always had the feeling that there was no better job for me at that time.

When I became a manager at Microsoft and was able to work to influence the work environment of the teams, my goal was for people to feel the same way. At Stone, a large part of my personal goal is also this: for every employee to feel they have the best job possible, with satisfied customers who choose the company for its efficient and affordable services. It's the greatest value I can extract from my professional experiences as a leader, and it's the yardstick I use to measure my performance.

---

56. The expression "for fun and profit" has been used in a multitude of book titles since the 1960s, from *Baby-Sitting for Fun and Profit* (1984), *Ancient Coins: How to Collect for Fun and Profit* (1964), to *Hunting Vampires for Fun and Profit* (2021), and in a series of books on computing, such as *Distributed systems for fun and profit* (2013).

I can't force people to think they're working in the best place. I can actively work to make it happen. Just as I have no direct control over the mood of the shareholders, nor do I have a lever to the number of clients the company attracts monthly, what I do control is hiring the right people, organizing the team, innovating, reducing technical debt, building scalable platforms, and dealing with incidents—all of which contribute to better products.

I've succeeded in my goals when I see that employees feel well rewarded, work with good tools, develop innovative technology, and have fun doing it. It may seem intangible, but you can feel the pulse of people daily. The leader needs to notice these signals and know where to act to correct the course of action. The environment may be excellent, but the development tools are terrible; the platforms may be great, but the people are not well rewarded, resulting in high turnover rates. The culture isn't collaborative, so that's what the leader needs to focus on. It may be that the platforms are not scalable, and this should be the main point of attention.

In the virtuous circle I've described, there is less staff turnover, and the technology team works more efficiently, which allows it to be leaner. Everyone has a well-defined contribution, delivering the platforms with the scale the company needs at low cost. An optimum point is then reached, in the spirit of Christopher Alexander and his "timeless way of building," where what is best for individuals and the company are not two conflicting components but synergistic ones.

Back to the beginning

I've had countless inspiring mentors throughout my career, and here, when I talk about synergies, I go back to the beginning. My first professional experience was under the command of a true Yoda,[57] Eugene "Beau" Shekita, who was incredibly careful with his team. It was very fortunate for a young person like me to start my career at IBM because, as a mature company with very structured management processes, it had a culture of giving a lot of attention to anyone coming in. Beau created a spirit of unity and collaboration among the team by establishing a routine of collective lunches, followed by hikes on the trail around the company's laboratory in California. According to him, the ritual was a way of preventing everyone from spending their days locked in their respective offices, which at the time were individual, without talking. Beau considered it essential that the team had the chance to exchange ideas and collaborate, and this conviction was what inspired me most at the time, even more than his enormous technical ability.

I also admire his ability to multiply talent. From the same group came Nadav Eiron, who was my leader at Google, and Vanja Josifovski, who I've already mentioned in earlier chapters and who is now CEO and

---

57. Yoda is the wise old Jedi master from the *Star Wars* series.

founder of an artificial intelligence company. It was a privilege to have Beau as my first leader in my professional life. Even more so following an academic mentor as dedicated and inspiring as Professor Lucena, from PUC-Rio.

My path crossed with that of Lucena[58] when I was still in high school. Enchanted by numbers, I was convinced I would take the math entrance exam. My aunt and uncle were Lucena's neighbors and advised me to talk to the professor, who had founded not only PUC-Rio's Computer Science Department but also the Padre Leonel Franca Foundation, which enabled a rich and long-lasting partnership between the university and industry. My parents, who were hoping that I would change my mind and decide to study engineering instead of mathematics (so as not to "starve" in the future), encouraged the meeting, and Lucena, who was already an expert in his field, not only agreed to receive me at PUC-Rio but also introduced me to a professor in the mathematics department, which only further cemented my idea of studying mathematics.

I only deviated from mathematics to engineering—much to my parents' relief—because I fell in love with computer science during my first algorithms course. I met Lucena again years later, towards the

---

58. More detailed information about the researcher's career and awards is available on the PUC-Rio website: https://www.inf.puc-rio.br/blog/professor/carlos-jose-pereira-de-lucena/ (accessed May 6, 2024).

final years of college, when I took a class with him. He didn't remember me, and I didn't want to allude to that meeting so as not to give the impression that I wanted some kind of favoritism from someone who was already an inspiration then. I was fortunate enough to be supervised by him for my master's and doctorate. The formal, serious man with completely white hair—he hadn't even turned 50 yet—became a mentor and friend. He was also a professor at the University of Waterloo, Canada, and I accompanied him three times to Waterloo during graduate school. At the living room table of his house in Teresópolis, I witnessed and participated in the creation of AulaNet, one of the first distance education platforms in Brazil. And it was from him that I learned the value of supervising people while giving them freedom to innovate.

During his post-doctorate at the IBM research laboratory, Lucena had been mentored by another great pioneer I've already mentioned here, Jean Paul Jacob. It was only years later that I became aware of their importance in the development of the technology industry in Brazil and the impact they had on me.

In terms of management, their approach taught me not to get so attached to details and to take a systemic view, focusing on the global impact. In a way, they both managed to distance themselves from the specific to move on to greater achievements.

It was this impulse that guided me—unplanned, I admit—in the search for large-scale impact throughout my career. If I had stuck to the area of distributed programming, close to the one I had dedicated myself to in graduate school, I wouldn't have chosen, when I joined IBM, to switch areas and work on databases. Perhaps unconsciously, since that time, I had chosen to be close to the best people, the "artists," and at IBM they were in the team that had created the first relational database.

Without the search for the greatest impact that Lucena instilled on me, I wouldn't have abandoned the area of search engines, in which I had become an expert, to work on the infrastructure of cloud computing services, a totally new subject for me, as I did when I joined Azure. And I wouldn't have accepted the challenge of bringing what I had learned about technology management over the years to Stone, a Brazilian fintech company.

Although what I do today is different from what Lucena did, the functions are similar in the sense that he acted like a point guard in basketball, distributing the balls and enabling others to do great things. When I think of Lucena, I immediately remember a persistent word in his vocabulary: "systematic." "Let's try to make a systematic model," I often heard, which meant **solving a problem in a generic and reproducible way**, which is more profound—and effective—than simply solving the problem.

A systematic model

Designing a systematic leadership framework is what I've tried to do throughout this book. I've tried to answer the question of how an engineering leader thinks—the platform mindset. In these pages, I've presented a systematic view of technology management, inspired by Lucena's approach and my own experiences. My goal is to fine-tune this management system so that it can enable the development of high-performance teams and generate business value through technology and innovation.

Computer science is a relatively new field that involves a lot of collaboration and interaction between people. Today, with platforms and abstractions such as cloud services, it's becoming easier to develop systems. Having a business idea and building a system to realize it is not rocket science. Almost anyone can build a prototype, even without much technical training. However, with the intentionality described here, this prototype will become a scalable business.

It must be scaled up with engineering processes, and architectural standards, tools, and platforms. Invariably, prototypes made by company founders need to be rethought at some point. That's where the importance of solid technical leaders comes in. They will question: Are all the security requirements being met? If users grow by 10x or 100x, can the system handle them? Are there mechanisms in place to combat fraud? Are all regulatory requirements being met? A group of young people making an app is one thing, building a system that will serve millions of people is quite another.

Experienced leaders' role is to encapsulate these standards and guidelines in tools, platforms, and culture so that knowledge can be passed on and reused automatically, thus achieving the desired multiplier effect.

That's why I was so excited about the chance to manage the 2,000 people in Stone's technology team. Disseminating the vision of a systematic management model could mean that more people will reflect on these practices, improve them, and teach them to others. Such multiplication has the potential to raise the level of engineering in Brazil, which in turn can positively affect socio-economic conditions in the country. As I'm not a politician, technology is my vehicle for improving people's lives.

When I started studying, I didn't think about any of this. All I knew was that I liked mathematics and was fascinated with programming. I also enjoyed working with people and was happy to work on complex projects involving several different teams. There is no other way to solve the most complex problems than to create collaborative teams, with engaged and talented employees. I then realized that I was good at organizing these teams and became aware of the potential this has in people's lives. Now I can give back all the training I received, to help train other people, and this book is part of that effort.

I'm very grateful for everything that has happened in my career. First doing a Ph.D., then working at Princeton and IBM, becoming a good developer, working in search engines, experiencing the creative and collaborative ebullience of Google at its peak, having the opportunity to lead a large organization at Microsoft. Things just kept happening. Looking back, one might even imagine that I planned every step. When I reflect on it myself, I'm amazed to realize that the evolution of my career has followed a coherent and rational path. What always guided me, however, was a desire to learn more, challenge myself, work on impactful projects, and why not, have fun.

I'm proud of creating environments where people can learn and feel inspired. Especially me, who didn't have the self-confidence even to read aloud at school, who thought I would never be able to work in a company like Microsoft, be a manager of a large organization, and speak in public to an audience of highly qualified professionals. One by one, the insecurities fell away with intense internal work, yes, but with the support of the people around me. Today, I have helped to bring down the insecurities of my mentees, inspiring those who lack confidence and are uncertain about the road ahead.

More than any project I've worked on, what stands out to me are the people I've worked with. I'm happy to have grown professionally unexpectedly, with the paths my career has taken. Looking back now, it was a systematic process, which I've tried to describe and dissect here. In my experience with yoga, I learned that you can't see the forest from inside, you must distance yourself from it. It was this search for a unifying framework that could be applicable and useful to others, that I sought in this report.

When I finished my graduate studies, I thought I would be a university professor in Brazil. My path went in another direction, but it took all those turns to realize that developing people is what really attracts me. I still want to do more to train professionals in Brazil, which I have been trying to do through my work at Stone, by supporting projects linked to education and, in a more informal way, by passing on my values in the day-to-day work interactions.

If I have contributed even a bit to forming a new generation of leaders with the mindset to build collaborative, diverse, innovative, and respectful work environments throughout my country and the world, I will have achieved my goal.

IMPORTANT POINTS ABOUT LEADERSHIP ATTRIBUTES

- Lead by example, including work-life balance.
- Have self-knowledge and capacity for self-criticism.
- Have good technical judgment.
- Build a team to complement your skills.
- Manage frustrations and team expectations.
- Be aware of the company mission.
- Strive for a systematic management model.
- Multiply knowledge and train new leaders.

Afterword

In today's rapidly evolving tech landscape, where the adage "every company is a software company" rings truer than ever, the AI revolution underscores the critical need for companies to understand how integrating technology management deeply across their entire organization is fundamental to their success. In that context, Marcus's insightful work, *Platform Mindset*, is an essential guide for navigating the intricacies of technical culture and organizational dynamics. This book transcends typical academic discourse, offering a narrative rich with authenticity and emotional depth. What sets this work apart is how he extracts valuable lessons from his personal experiences, conveying guidance on technology management through engaging storytelling.

While working alongside Marcus while he was in Microsoft Azure, I witnessed his impressive journey from individual contributor to leader of a diverse team. His significant contribution to Azure's evolution is a testament to the thoughtful way he studied how to be an effective individual contributor, and then a technical manager that designed his organization and curated its culture so that his team could thrive and deliver impact. He earned the Technical Fellow title, Microsoft's highest technical distinction, based on his ability to identify the most important opportunities, to propose approaches to take advantage of them, and to bring the rest of the organization along with his vision. We tried very hard to retain him when he decided to leave Microsoft for Stone, not only because of his impact, but the positive influence he had on everyone that worked for him and with him.

This book stands out in its comprehensive exploration of technology roles, success factors, and organizational structures that foster innovation. It covers the nuances of technical leadership, the nurturing of emerging talent, and the pathways for technical career advancement. Throughout, Marcus weaves in the wisdom gained from his experiences, both triumphs and setbacks, offering readers a unique blend of practical advice and hard-won insights. He shares profound challenges, including the loss of his wife and the turmoil of divorce, not merely as anecdotes, but as learning opportunities from which he draws insights into resilience, adaptability, and empathetic leadership in the tech world.

As we stand on the brink of the AI revolution, Marcus's perspective provides a balanced view of the potential challenges and opportunities that organizations may encounter. His work affirms that mastering technology management extends beyond tools and processes—it's fundamentally about understanding and guiding the people who harness them, a lesson he clearly draws from his own journey through various roles and life experiences. It's an essential read for anyone aiming to navigate the complex world of technology management with skill and insight, learning from both Marcus's successes and the valuable lessons he's gleaned from life's challenges.

Mark Russinovich
CTO and Technical Fellow, Microsoft Azure

With a little help from my friends

I mention throughout the book several people who were extremely important in my life and career, and who helped me define the management concepts described here. However, I feel I attended a concert where the artist sang many songs I like but left out many others. Despite omitting the names and stories of several significant people to me, I want to express my sincere thanks to everyone I've worked with throughout my career.

This book would not have been possible without the total and complete dedication of journalists Carolina Schwartz and Fernanda Ravagnani, who, in countless hours of conversation and discussion, challenged all the concepts and ideas, and helped me to synthesize my management principles in the best possible way, making them clear to non-technical readers.

I would also like to thank my colleagues at Stone: Fernanda Teich, Leonardo Guineli, Mariana Rigoni, Fabio Kapitanovas, and Anderson Nielson, who provided valuable feedback to earlier drafts. My friend Paulo Caroli also deserves thanks for his support during the writing and publishing process. The team from Citadel for the tremendous work on the

Portuguese version of this book. Greg Shaw, Myles Thompson, Karen Bowen, Barbara Byrne, and the team from 8080 Books for the support and hard work on this edition.

The initiative to author this book arose shortly after the passing of my friend Luiz André Barroso. In the last months of his life, Luiz was part of Stone's board of directors, and I was very close to him, after several years of little contact while I was at Microsoft. His death, so unexpected, deeply affected me and made me take this project off the shelf.

As I mention throughout the book, my family provided me with an extremely favorable starting point. My parents, Vanessa and Julio, always valued education and always supported my sister, Tatiana, and I, believing in our potential and encouraging us to do our best. I owe my career to them.

Finally, being a father has made me a better person and a better professional. I have learned and continue to learn a lot from my daughters, Ayesha and Shanti. Participating in their upbringing and growth, and witnessing how they are becoming incredible people, has been one of the greatest joys of my life. Many of the technology concepts described here apply to the context of life itself, however, I sometimes feel it is easier to be a good manager at work than a parent who gets it 100% right at home. I continue striving to improve every day. Thank you, Ayesha and Shanti.

About the author

Marcus Fontoura is the CTO at Stone, a leading provider of financial technology and software solutions. Previously he worked at Microsoft as a technical fellow and architect for Azure Compute. He also led the Azure efficiency team. He worked on several projects including container allocator, power management, and Resource Central, a machine-learning infrastructure for resource management. Prior to joining Azure, Marcus worked on the production infrastructure for Bing and several Bing Ads projects, including ad infrastructure and relevance.

Before Microsoft, he was a Staff Research Scientist at Google, working in the Search Infrastructure team. His focus was on the serving systems powering Google.com search.

Before joining Google, he was a Principal Research Scientist at Yahoo Research working on several projects in computational advertising. He's also worked as the architect for a large-scale software platform for indexing and content serving, which is used in several of Yahoo's display and textual advertising systems.

Before Yahoo, he worked as a Research Staff Member at the IBM Almaden Research Center, where he co-developed a query processor for XPath queries over XML streams. This was one of the critical components for implementing the XML data type in the IBM

DB2 Relational Database System. In another project at IBM, he was one of the key researchers developing an Enterprise Search Engine. This project resulted in a new software product for IBM—the IBM OmniFind Enterprise Search. For this work, he was awarded an IBM Outstanding Technical Achievement Award, with the notation for developing a new generation of IBM search technology and its deployment on w3.ibm.com.

Marcus holds a Ph.D. from the Pontifical Catholic University of Rio de Janeiro, a joint program with the Computer Systems Group, University of Waterloo, Canada. His Ph.D. work was in object-oriented design and software architecture. The main contributions from his Ph.D. thesis have been condensed in the book *The UML Profile for Framework Architectures*, published by Addison-Wesley in 2001. He was also a postdoctoral researcher in the Computer Science Department at Princeton University.

Other titles from 8080 Books

No Prize for Pessimism by Sam Schillace

BOOKS

Check for new titles on our website:

https://unlocked.microsoft.com/8080-books/